KNOWLEDGE ORGANISATION, INFORMATION PROCESSING AND RETRIEVAL

Knowledge Organisation, Information Processing and Retrieval

Editor

Fernand Qunicy

scitus
academics

Knowledge Organisation, Information Processing and Retrieval

Edited by **Fernand Qunicy**

ISBN: 978-1-68117-560-7

Library of Congress Control Number: 2016942974

SCITUS Academics LLC,
www.scitusacademics.com
Box No. 4766, 616 Corporate Way,
Suite 2, Valley Cottage,
NY 10989

Notice

Printed in United States of America on Acid Free Paper

PREFACE

Literally Knowledge Organisation is the organisation of knowledge. This organisation is made in order to facilitate the use of documents or recorded knowledge (or other units). Information or knowledge management is essentially a four-step process that includes capturing, organising, refining and disseminating information. The process begins with the capturing of information relevant to the activities and interests of people in an organisation. Such information may be available in a variety of sources, forms and formats. Once information is captured, it needs to be organised using a number of techniques that include cataloguing and indexing, retrieving, filtering, ranking, and so on. Efficiency of an information access system depends largely on the proper organisation of information. Information processing is the acquisition, recording, organisation, retrieval, display, and dissemination of information. Over the years, information science researchers and practitioners have developed and used various techniques for organising information resources of different types. However, off late, with the introduction of information and communication technologies several sophisticated systems have been developed for organisation of information. Libraries and information services have a long history of using various tools for organising information resources.

Knowledge Organisation, Information Processing and Retrieval focuses on the various bibliographic and information retrieval tools and techniques used for information organisation, a key activity in a knowledge management process. The book begins with the existing debate on the very concept of knowledge management, and looks at some recent papers and arguments on this issue. It then briefly discusses how some projects over the past decade or so have used various traditional bibliographic organisation tools for providing access to electronic resources.

TABLE OF CONTENTS

1

SYNTACTIC SENTENCE COMPRESSION IN THE BIOMEDICAL DOMAIN: FACILITATING ACCESS TO RELATED ARTICLES

Jimmy Lin[1], W. John Wilbur[2]

[1]College of Information Studies, University of Maryland, College Park, MD, USA

[2]National Center for Biotechnology Information, National Library of Medicine, Bethesda, MD, USA

ABSTRACT

We explore a syntactic approach to sentence compression in the biomedical domain, grounded in the context of result presentation for related article search in the PubMed search engine. By automatically trimming inessential fragments of article titles, a system can effectively display more results in the same amount of space. Our implemented prototype operates by applying a sequence of syntactic trimming rules over the parse trees of article titles. Two separate studies were conducted using a corpus of manually compressed examples from MEDLINE: an automatic evaluation using BLEU and a summative evaluation involving human assessors. Experiments show that a syntactic approach to sentence compression is effective in the biomedical domain and that the presentation of compressed article titles supports accurate "interest judgments", decisions by users as to whether an article is worth examining in more detail.

KEYWORDS

Sentence compression; Extrinsic evaluation; PubMed MEDLINE; Genomics IR

1. INTRODUCTION

Sentence compression has previously been identified as a key component in document summarization. Indeed, the ability to convey the substance of a piece of text, but in a smaller amount of space, is one hallmark of a good summarization system. Although both syntactic and statistical approaches have been employed to tackle this problem, previous attempts largely focus on newswire text. This work differs from previous studies of sentence compression in two important ways: First, we explore an application of syntactic trimming techniques in the biomedical domain. Other than brief mentions by Ruch et al. (2003) and Lu et al. (2006) in the context of GeneRIF extraction, this work represents the first systematic attempt, to our knowledge, at tackling sentence compression in this highly-specialized domain. Second, we couch sentence compression within the context of result presentation in an information retrieval task. This framing of the problem provides an extrinsic, task-based evaluation grounded in real-world user scenarios.

We present a sentence compression algorithm for article titles from MEDLINE based on syntactic trimming rules that operate over parse trees. This approach was adopted due to its proven effectiveness in previous summarization tasks and the paucity of training data in the biomedical domain. Intrinsic evaluations show that trimmed titles generated by our system are competitive with manually-compressed gold standards, both in terms of automatic metrics (BLEU) and human judgments of content and fluency. In addition, we evaluated the ability of our compressed titles to facilitate "interest judgments"—the decision by a user regarding whether or not an article is worth examining in response to an information need. We found little difference between original and automatically compressed titles, indicating that from a task viewpoint, our system can effectively support users' decisions while reducing the amount of text they must read.

This article is organized as follows: We begin by describing the task model that underlies our explorations in Sect. "Motivation". Previous work is reviewed in Sect. "Related work". Efforts to develop appropriate data resources are outlined in Sect. "Resource development". The syntactic trimming algorithm is detailed in Sect. "A syntactic approach to compression". Evaluation is broken up into two sections: Sect. "Automatic evaluation" covers automatic evaluations, while Sect. "Manual evaluation" covers manual evaluations. We discuss the significance of our work in Sect. "Discussion" and future plans in Sect. "Future work" before concluding.

2. MOTIVATION

Our work is situated in the context of the PubMed search engine,1 a freely-accessible gateway to the MEDLINE bibliographic database maintained by the US National Library of Medicine (NLM). This database is viewed by medical professionals, biomedical researchers, and many other users as the authoritative source of information related to the health sciences. MEDLINE contains over 15 million references to articles from approximately 5,000 journals in 37 languages, dating back to the 1960s. In 2006, over 623,000 new citations were added to the database, and it currently grows at a rate of 2–4,000 citations daily. The subject scope of MEDLINE is biomedicine and health, broadly defined to encompass those areas of the life sciences, behavioral sciences, chemical sciences, and bioengineering needed by health professionals and others engaged in basic research and clinical care, public health, health policy development, or related educational activities. MEDLINE also covers life sciences vital to biomedical practitioners, researchers, and educators, including aspects of biology, environmental science, marine biology, plant and animal science as well as biophysics and chemistry.2

Each MEDLINE citation includes basic metadata information such as the title of the article, name of the authors, name of the publication, publication type, date of publication, language, etc. Of the entries added over the last decade or so, approximately 79% have English abstracts written by the authors of the articles.

PubMed is a freely-accessible Web search engine that provides access to the MEDLINE database, developed by the National Center for Biotechnology Information (NCBI) at NLM. The system provides an array of query operators that allow users to query in specific data fields (title, author, etc.) and leverage Medical Subject Headings (MeSH), drawn from NLM's controlled vocabulary thesaurus. MeSH terms are assigned manually by trained human indexers with the assistance of automated systems (Aronson et al. 2004).

A recently-revised functionality in PubMed is the "Related Links" feature. When the user examines a MEDLINE abstract, the right panel of the browser is automatically populated with titles of articles that may also be of interest, as determined by a probabilistic content similarity algorithm (Wilbur 2005)—in short, PubMed implicitly issues a query for related articles whenever a user pulls up a MEDLINE abstract to examine in detail. The goal of this feature is to unobtrusively suggest other interesting citations to facilitate knowledge discovery. The screenshot in Figure 1shows the arrangement of the PubMed search interface.

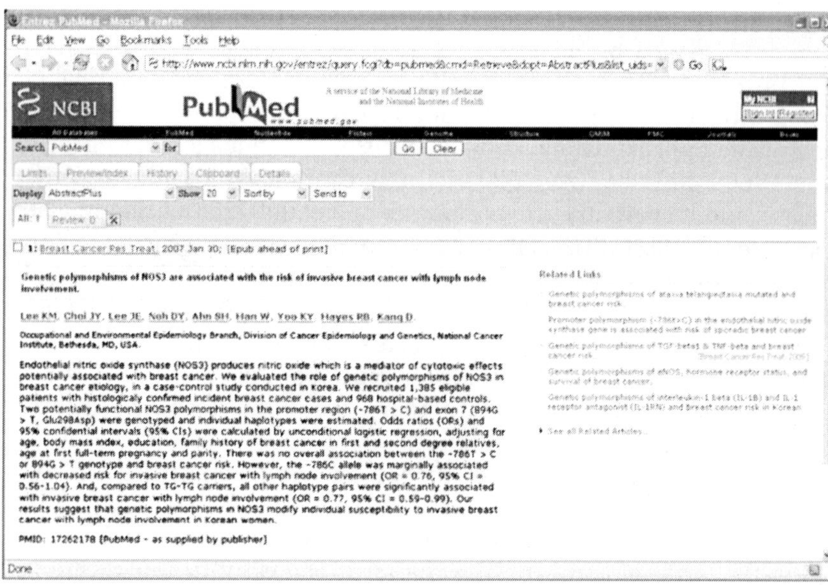

Figure 1. Typical screenshot of the PubMed search engine as the user examines an abstract. The "Related Links" panel on the right is populated with titles of articles that may also be of interest. Other currently empty space is reserved for future functionality

Article titles serve as document surrogates for presentation in the "Related Links" panel. Other space in the current PubMed interface is reserved for additional features that have yet to be deployed. Since the screen area available to this feature is fixed, we are faced with a tradeoff between quality and quantity: either show more related articles in less detail or show fewer related articles in more detail. In the simple case, this tradeoff can be implemented by controlling the amount of associated metadata that is displayed and truncating the title based on a fixed length quota. This work explores a more sophisticated solution based on linguistic analysis—we hypothesize that sentence compression techniques can potentially deliver the best of both worlds: by shortening article titles in a linguistically meaningful way, the interface could deliver much of the same substance in a smaller amount of space.

Ideally, the output of a compression algorithm should be both indicative and informative—properties often discussed in the context of document summarization (Afantenos et al. 2005). Indicative summaries suggest the content of an underlying information object without necessarily giving away details—often the aim is to entice users, or at least alert them to the presence of a particular information object. Movie trailers and book jackets are good examples. In contrast, an informative summary is meant to represent (and sometimes replace) the original object. Take the case of summarizing news articles: indicative summaries might mention the entities involved in a particular event, but not actually say *what happened*. An informative summary might focus on what happened, but neglect to mention the roster of participants. See (Kan et al. 2001) for an example of an attempt to introduce this distinction into summarization systems. In our application, document surrogates for related articles (displayed in the "Related Links" panel) should ideally satisfy both properties—be indicative as to capture users' interest and be informative in order to convey sufficient substance.

Ultimately, the related links feature in PubMed is designed to guide users to other articles of interest—to support information seeking or knowledge discovery. This underlying task model guides our work and provides a framework for extrinsic evaluation. The goal of the document surrogates (i.e., compressed article titles) is to facilitate interest judgments, or user decisions on whether or not to examine a particular citation. We can realistically measure the effectiveness of a compression algorithm by its ability to support such decisions.

In this context, the notion of "interest judgment" differs from the more traditional judgment of relevance that forms the basis of most retrieval applications. Ultimately, retrieval systems aim to deliver information that addresses users' information needs. However, relevance is a multi-faceted consideration that takes into account a multitude of factors—see (Mizzaro 1998) as a starting point into the rich body of literature on relevance. We do not believe it is possible to assess an article's relevance from only the title (or any short surrogate), and hence it would not be meaningful to examine relevance directly in our task context. However, an important intermediate step is the decision to examine an abstract in more detail—which we call an interest judgment. Such a decision will then cause a user to bring up more details about the article (abstract text, authors, and other metadata) in order to make a more informed decision about relevance. The related links feature in PubMed is exactly designed to elicit such interest.

Furthermore, our definition of "interest" opens the door to different types of relations beyond relevance—a citation may be interesting, not because it is potentially relevant to the present information need, but because it raises questions that the user may not have previously considered. These types of serendipitous connections underlie many significant breakthroughs in the life sciences, and PubMed aspires to assist in this process of knowledge discovery by drawing links where none previously existed.

3. RELATED WORK

Our approach to sentence compression is most similar to the work of Zajic et al. (2004) in that both use a series of linguistically-motivated trimming rules to remove inessential fragments from the parse tree of a sentence—additional details can be found in (Dorr et al. 2003; Zajic et al. 2007, in press). This approach has proven to be highly effective at generating very short summaries of single newswire documents (i.e., the headline generation task), as evidenced by its performance at the DUC 2004 summarization evaluation. Topiary, the University of Maryland's system which integrates "parse-and-trim" techniques with topic term extraction, was among the highest scoring systems for all tasks on all measures—in some cases, even beating the performance of humans in terms of automatic metrics. Other systems that make use of similar techniques include (Mani et al. 1999; Jing 2000), and more recently, (Blair-Goldensohn et al. 2004; Conroy et al. 2005). In our approach, sentence compression is achieved by removing elements—no attempt is made to reorder material within a sentence. Since our task does not involve multiple sentences, there is no opportunity to generate output that combines fragments from different sources, for example, including one sentence as a relative clause inside another (Mani et al. 1999). Thus, we conceive of sentence compression solely as the task of selecting sentential elements (words, phrases, clauses, etc.) to remove.

Sentence compression has also been tackled with supervised machine learning techniques using a noisy-channel model. Verbose text can be viewed as the output of passing the original compressed text through a "noisy channel" that inserts additional inessential content. Given the verbose text, the system's task is to reconstruct the original message. The problem can be modeled in terms of simple word-level features, as in (Banko et al. 2000), or in terms of parse tree structures, as in (Knight and Marcu 2000; Turner and Charniak 2005). One downside of these statistical approaches is the need for annotated training data to learn model parameters. On the other hand, since trimming rules are able to exploit human linguistic insight, far less data is required for system development. Nevertheless, both methods can be viewed as complementary.

Another approach to generating very short summaries is to extract a list of topic descriptors indicative of content; examples include (Bergler et al. 2003; Zhou and Hovy 2003; Wang et al. 2005). The output of such techniques consists of, for the most part, noun phrases—as such, they are useful for telling a user what the important entities are, but less useful for conveying what actually happened. In other words, system output is indicative, but often not informative.

Although document summarization techniques have principally been applied to newswire text, there is a body of research that deals specifically with the summarization of medical documents—see (Afantenos et al. 2005) for a survey. A noteworthy example is PERSIVAL (McKeown et al. 2003; Elhadad et al. 2005), which leverages patient records to generate personalized summaries. In the genomics domain, automatic summarization techniques have also been applied to extracting GeneRIFs, concise phrases describing the function of genes (Ruch et al.2003; Ling et al. 2006; Lu et al. 2006). In particular, Ruch et al. (2003) and Lu et al. (2006) both briefly mention methods for removing inessential elements from GeneRIFs, which are similar in spirit to our sentence compression techniques. In comparison to these cited articles, both on summarization of newswire and biomedical text, what sets our work apart is a focus on sentence compression as a tool to facilitate knowledge discovery in the context of an information retrieval system.

An important part of summarization research focuses on methodologies for evaluating system output, which can be broadly classified into two categories. In an *intrinsic* evaluation, system output is directly evaluated in terms of some set of norms—for example, fluency (Minel et al. 1997), coverage of key ideas (Paice1990; Brandow et al. 1995), or similarity to an "ideal" summary (Kupiec et al.1995). In particular, the last criterion has been operationalized in ROUGE (Lin and Hovy 2003), an automated metric that compares system output to a number of human-generated "reference" summaries. The primary difficulty, however, lies in establishing an ideal reference (or a set of such texts)—summaries are generated for different purposes, and the human-centric nature of the task means that there is more than one "correct answer". Operationally, this results in low interannotator agreement on tasks such as sentence extraction (Salton et al. 1997).

In contrast to *intrinsic* evaluations, *extrinsic* evaluations attempt to measure how summarization impacts some other task. Developing realistic usage scenarios is challenging, but often the "goodness" of a summary can only be meaningfully operationalized in its "usefulness" for a particular task. One might, for example, measure how summaries impact question answering (Morris et al. 1992; Mani et al. 2002) or relevance judgments (Dorr et al. 2005). One possible hypothesis is that summaries allow users to make quicker decisions (since they have to read less), without compromising the quality of those decisions. Along these lines, our work is grounded in an information retrieval task, which allows us to assess the potential real-world impact of our sentence compression techniques.

4. RESOURCE DEVELOPMENT

Since we are not aware of any existing resources for the sentence compression task in the biomedical domain, we devoted significant effort to creating a corpus of annotated examples. Our collection consists of article titles that have been manually shortened by domain experts. Instead of randomly sampling citations from the MEDLINE database, we leveraged the test collection developed from the TREC 2005 genomics track (Hersh et al. 2005), which fits well with our task model (finding articles of interest in the context of an ongoing search for information).

One salient feature of the TREC 2005 genomics track evaluation is its use of generic topic templates (GTTs) to capture users' information needs, instead of the typical free-text title, description, and narrative combinations used in other *ad hoc*retrieval tasks. The GTTs consist of semantic types, such as genes and diseases, that are embedded in common genomics-related information needs, as determined from interviews with real biologists. In total, five templates were developed, with 10 fully-instantiated topics for each—examples are shown in Figure2. Note that in some cases, the actual topics deviated slightly from the template structure (in order to accommodate real requests).

Template #1: Information describing standard [methods or protocols] for doing some sort of experiment or procedure.
methods or protocols: How to "open up" a cell through a process called "electroporation"

Template #2: Information describing the role(s) of a [gene] involved in a [disease].
gene: Interferon-beta
disease: Multiple Sclerosis

Template #3: Information describing the role of a [gene] in a specific [biological process].
gene: nucleoside diphosphate kinase (NM23)
biological process: tumor progression

Template #4: Information describing interactions (e.g., promote, suppress, inhibit, etc.) between two or more [genes] in the [function of an organ] or in a [disease].
genes: CFTR and Sec61
function of an organ: degradation of CFTR
disease: which leads to cystic fibrosis

Template #5: Information describing one or more [mutations] of a given [gene] and its [biological impact or role].
gene with mutation: BRCA1 185delAG mutation
biological impact: role in ovarian cancer

Figure 2. Templates and sample instantiations from the TREC 2005 genomics track

The genomics track employed a 10-year subset of the MEDLINE database (1994–2003), which totals 4.6 million citations, or approximately a third of the size of the entire MEDLINE database at the time it was collected in 2004. Each citation is identified by a unique pmid. In total, 32 groups submitted 59 runs to the task (both manual and automatic), which insured a rich, diverse pool of results. Relevance judgments were provided by an undergraduate student and a Ph.D. researcher in biology.

First, we randomly sampled 200 titles from the known list of relevant citations, and another 200 titles from the known list of irrelevant citations. These were then merged and randomized, producing a total of 400 titles, half of which were relevant according to the original assessors in the TREC task. We adopted this sampling process in order to obtain a good balance—a truly random sampling of the citation pool would yield a much larger fraction of irrelevant documents.

Next, these titles were presented to two human annotators—both were subject domain experts otherwise uninvolved with the project. Throughout this paper, we will refer to these annotators as "Lo" (Ph.D. in bioinformatics, B.S. in molecular biology) and "Ly" (Ph.D. in human genetics). They were provided both the original title and the information need that the citation was retrieved to address (i.e., the template with concept instantiations). Both annotators were asked to generate a compressed version of the title by removing unimportant elements from the full title. After this was done, the 400 annotated pairs were divided into a development set and a held-out test set.

We attempted to align the resource development process with our task model as much as possible. To start, the TREC genomics track employed a subset of the MEDLINE database, which gave us a degree of confidence that findings could be directly applied to PubMed. Human generation of the compressed titles occurred in the context of information needs, much like the task setup in related article search (i.e., browsing the "Related Links" panel in PubMed). Furthermore, the needs are those typical of a particular user population, since they were generalized from interviews with real biomedical researchers. We believe that this collection encapsulates the original end-to-end task with great fidelity, thus enabling the results of laboratory experiments to be applicable in real-world environments.

Characteristics of our annotated data are shown in Table 1. We show the average compression ratio and the average length reduction (with standard deviation) both in terms of characters and words. Compression ratio is computed as the length of the compressed sentence divided by the length of the original sentence, averaged across the entire data set of 200 sentences. Thus, the smaller the number, the shorter the output is. We note that there appear to be more opportunities for compression in the development set (given the lower compression ratios). Abstract titles in the development set averaged 102.3 characters, or 13.34 words; abstract titles in the test set averaged 104.0 characters, or 13.60 words.

Table 1. Characteristics of the human-generated compressed abstract titles

Annotator	Characters		Words	
	Ratio	Reduction	Ratio	Reduction
Development set				
Lo	0.564	46.4 ± 28.4	0.543	6.34 ± 3.81
Ly	0.575	48.0 ± 34.9	0.567	6.40 ± 4.61
Test set				
Lo	0.601	43.6 ± 26.5	0.587	5.99 ± 3.77
Ly	0.652	39.0 ± 30.6	0.644	5.25 ± 4.12

The table shows compression ratio and average length reduction (with standard deviation) at the character and word levels: development set on top and test set on bottom. Length of complete abstract titles: 102.3 characters, 13.34 words (for development set) and 104.0 characters, 13.60 words (for test set)

5. A SYNTACTIC APPROACH TO COMPRESSION

We adopt a syntactic approach to sentence compression through the use of linguistically-motivated trimming rules that remove fragments of parse trees. Our work employs the Stanford Parser (Klein and Manning 2003).3 Although it was not originally designed to parse text in the biomedical domain, experimental results show that a syntactic approach is nevertheless effective for compressing abstract titles in the biomedical domain (more on this in Sect. "Discussion"). The techniques described here are not tied to a particular parser and will function with any system that utilizes the Penn Treebank conventions.

From the development set, we came up with seven linguistically-motivated rules—many of these are similar to those discussed in (Zajic et al. 2004). These rules are described below, arranged roughly in increasing order of complexity. An example of each is shown in Table 2; those examples illustrate the application of each rule in isolation.

- *Subtitles:* Subtitles, denoted with a colon or consecutive dashes, are removed. This is accomplished with simple regular expression patterns prior to parsing. Examination of titles from the development set suggests that the recognition of colons and consecutive dashes is sufficient to identify subtitles.

- *Determiners (DT):* Determiners are removed. All terminals in the parse tree assigned the part-of-speech tag DT are considered determiners.

- *Participial and Gerund phrases (VBG):* Participial and gerund phrases are removed. These phrases are recognized by traversing the parse tree and identifying VPs headed by tokens tagged as VBG, i.e., the structure $[_{VP}[_{VBG}X]$. Resulting empty nodes (e.g., dangling prepositions) are also removed.

- *Serial PPs:* In a sequence of prepositional phrases (sharing a common ancestor), all but the first are removed. Prepositional phrases are recognized by traversing the parse tree and identifying constituents with the structure $[_{PP}[_{IN}X]\cdots]$. See Sect. "Discussion" for a detailed discussion regarding prepositional phrase attachment issues with the Stanford Parser.

- *Nested PPs:* Any prepositional phrase that is embedded three or more levels deep is removed. As with the previous rule, PPs are identified by traversing the parse tree and identifying structures of the following form: $[_{PP}[_{IN}X]\cdots]$ See also Sect. "Discussion" for issues related to prepositional phrase attachment.

- *Conjoined NPs:* For conjoined noun phrases, the second conjunct is removed. This rule is motivated by Zajic et al. (2007, in press), who observed that the first conjunct in a $[_{NP}[_{NP}\cdots]$ and$_{CC}[_{NP}\cdots]$ structure is often more important. Noun phrases containing conjoined adjectives are left untouched.

- *Simple NPs:* Adjectives in NPs are removed, unless their heads are "lightweight". For example, contrast "a specific$_{JJ}$promoter$_{NP}$ nucleosome$_{NP}$" and "the epsilon-globin$_{JJ}$ gene$_{NP}$": whereas the adjective can be removed without significantly affecting content in the first case, the token tagged as adjective ("epsilon-globin") actually conveys most of the content in the second NP. We term the head (rightmost) noun lightweight in the second case and approximate its "weight" using inverse document frequency (*idf*), a commonly-employed measure in information retrieval. Thus, a threshold associated with this rule specifies the weight of a head noun above which modifying adjectives are removed.

Table 2. Example of all trimming rules. All rules are applied in isolation

Rule	Example
Subtitle	Structure and dynamics of the GABA binding pocket~~: A narrowing cleft that constricts during activation.~~ (PMID: 11150321)
DT	~~The~~ effectiveness of ~~a~~ peripatetic allergy nurse practitioner service in managing atopic allergy in general practice — ~~a~~ pilot study. (PMID: 10779298)
VBG	Mutation analysis in a small cohort of New Zealand patients [$_{VBG}$ ~~originating from the United Kingdom~~] demonstrates genetic heterogeneity in familial hypercholesterolemia. (PMID: 11040093)
Serial PP	Electroporation-mediated interleukin-12 gene therapy [$_{PP}$ for hepatocellular carcinoma] [$_{PP}$ ~~in the mice model~~]. (PMID: 11221826)
Nested PP	Semiquantitative immunoblot analysis [$_{PP}$ of nm23-H1 and -H2 isoforms [$_{PP}$ ~~in adenocarcinomas of the lung~~]]: prognostic significance. (PMID: 10792783)
Conjoined NP	[$_{NP}$ [$_{NP}$ Extraction] ~~and~~$_{CC}$ [$_{NP}$ ~~isolation~~] of linear alkylbenzenesulfonate] ~~and~~$_{CC}$ [$_{NP}$ ~~its sulfophenylcarboxylic acid metabolites~~] from fish samples. (PMID: 10575968)
Simple NP	Acetylation of [$_{NP}$ a ~~specific~~$_{JJ}$ promoter$_{NP}$ nucleosome$_{NP}$] accompanies activation of [$_{NP}$ the epsilon-globin$_{JJ}$ gene$_{NP}$]. (PMID: 11158302)

The above trimming rules were formulated after examining sentence pairs in the development set and recognizing opportunities to remove inessential fragments of titles based on syntactic structure. This was not accomplished in a systematic way, and no doubt there are other opportunities that can be exploited. Nevertheless, as our evaluations show, these rules provide the basis for an effective sentence compression algorithm. The order in which the trimming rules are sequentially applied in our final implementation was guided by the evaluation of individual rules, described in the next section.

6. AUTOMATIC EVALUATION

Although end-to-end task-based evaluations provide the best method for assessing information systems, the large amount of manual effort typically required for such evaluations precludes using them for system development. In human language technologies, researchers often employ a paradigm based on automatic metrics for system development, capped with a summative evaluation. Our work follows in this model. This section describes a series of experiments that characterize the effectiveness of our syntactic trimming techniques using automated methods. A manual evaluation is detailed in Sect. "Manual Evaluation".

7. EVALUATION METHODOLOGY

Automatic evaluation methods are attractive because they enable quick experimental turnaround, thereby facilitating rapid exploration of the solution space. An easily quantifiable performance metric provides researchers with an objective function over which they can optimize. Once an automatic metric has been validated—that is, demonstrated to correlate with human preferences—the measure can be exploited for system development.

In the language processing community, researchers have developed a family of automatic metrics based on the idea of comparing system output to one or more human-generated references. Similarity to these "gold standards", according to different content overlap metrics, can quantify the quality of system output—this represents a well-established evaluation methodology in the language processing community. Two such commonly-used metrics are BLEU (Papineni et al. 2002) for machine translation and ROUGE (Lin and Hovy 2003) for document summarization. Both rely on computing *n*-gram overlap between system output and human references (manually-translated sentences and human-generated summaries, respectively), but the details differ. In general, BLEU is a precision-oriented metric that places heavy emphasis on fluency, i.e., checking to make sure that the system output is "good English". This is important in evaluating machine translations since automated systems have a tendency to produced garbled sentences. ROUGE, a metric developed for document summarization tasks, on the other hand, focuses on recall of content, i.e., the presence or absence of certain topic terms. Because most modern document summarization systems are extractive, generation of disfluent output is not as severe a problem.4 Thus, summarization metrics emphasize the inclusion of key facts or concepts in the system output, as measured in terms of *n*-gram overlap.

Given these considerations, we decided that BLEU is the more appropriate metric for our sentence compression task—primarily because we are concerned with the fluency of system output. Conceptually, sentence compression can be viewed as "translating" from verbose English into succinct English. According to previous work (Lin 2004), ROUGE-1 recall correlates best with human judgments on the headline generation task—constructing a very short summary (less than 75 characters) from a single newspaper article. This is the closest summarization analog to our task, but we believe that ROUGE-1 is not an appropriate metric. The measure focuses on unigram overlap, i.e., the presence of content words in system output, and is not sensitive to fluency considerations; a grammatical sentence and a random sequence of the same words would receive identical scores. This characteristic does not fit with our desire to assess the grammatical correctness of system output, which is an important concern since removing certain portions of the parse tree may yield ungrammatical output. BLEU, on the other hand, is better able to model fluent English text since it takes into account *n*-grams of different lengths. Although the metric considers only the surface properties of machine output and lacks even a rudimentary model of syntax and semantics, it has proven highly effective in guiding research in machine translation. Despite its deficiencies, BLEUhas enabled rapid progress in translation technology over the past few years.

To get a sense of how much the assessors agreed with each other, we computed BLEU scores, using one as the "system output" and the other as the reference. These results are shown in Table 3. Note that these values are lower than many of the BLEU scores reported for our compression algorithm because only one set of references is used; in all other experiments, system output is evaluated against both sets of human references.

8. APPLICATION OF INDIVIDUAL RULES

We first examined each rule in isolation. Table 4 shows results for all rules expect for the simple NP rule, which requires an additional parameter and is therefore discussed separately. For each rule, we note the number of titles that triggered the rule (out of 200 in each set) in the second column of the table. This value quantifies the prevalence of each phenomena. The next four columns show average compression ratio and average length reduction (with standard deviation), both at the character and word levels. These values are computed over affected titles only (the set of abstract titles that triggered the rule). The BLEU scores were computed across the entire data sets (all 200 sentences), using both "Lo" and "Ly" as the references.

Table 3. BLEU scores of one annotator's output using the other as a reference, on both the development set and the test set

Output	Reference	Dev	Test
Lo	Ly	0.510	0.550
Ly	Lo	0.513	0.544

These values characterize the agreement between the two human annotators

How is one supposed to interpret these results? The effectiveness of each rule is quantified in two ways: the amount of compression achieved and how "good" the resulting output is (compared to human-generated references using the BLEUmetric). Thus, each rule represents a tradeoff along these two dimensions. Leaving the abstract titles untrimmed (the first row in Table 4) represents a baseline. Not surprisingly, application of the trimming rules in most cases raises the BLEU score, indicating that the results are closer to the references than the original full title in terms of *n*-gram content. In most cases, gains observed in the development set carried over to the held-out test set, although the magnitude of the improvements were smaller (but recall that the human-annotated gold standards suggest fewer opportunities for trimming in the test data). In general, the scope of the rules (i.e., number of affected titles) and the degree of compression were comparable.

Table 4. Compression effectiveness of each rule in isolation with the exception of the simple NP rule

Rule	#	Characters		Words		BLEU
		Ratio	Reduction	Ratio	Reduction	
Development set						
Untrimmed	–	1.000	0.00	1.000	0.00	0.538
Subtitle	39	0.525	49.92 ± 27.36	0.518	6.54 ± 3.61	0.579
DT	103	0.955	4.64 ± 2.57	0.899	1.40 ± 0.66	0.555
VBG	20	0.584	47.20 ± 23.56	0.573	6.40 ± 3.10	0.541
Serial PP	40	0.587	39.85 ± 21.78	0.564	5.70 ± 3.12	0.527
Nested PP	41	0.748	32.85 ± 23.07	0.723	4.76 ± 2.99	0.540
Conjoined NP	73	0.722	29.30 ± 17.89	0.703	4.10 ± 2.53	0.514
Test set						
Untrimmed	–	1.000	0.00	1.000	0.00	0.593
Subtitle	38	0.510	53.45 ± 29.83	0.529	6.82 ± 4.20	0.619
DT	109	0.956	4.61 ± 2.68	0.907	1.34 ± 0.74	0.600
VBG	26	0.656	36.35 ± 29.24	0.683	4.31 ± 3.98	0.586
Serial PP	42	0.688	34.86 ± 28.18	0.670	4.93 ± 3.93	0.596
Nested PP	36	0.778	28.50 ± 20.16	0.759	4.22 ± 3.05	0.591
Conjoined NP	54	0.763	25.93 ± 16.22	0.753	3.57 ± 2.44	0.575

The second column shows number of abstract titles (out of 200) that triggered the rule. The last column shows the BLEU score of compressed output. Middle columns show average compression ratio and average length reduction (with standard deviation), in terms of characters and words. Data from development set shown on top, and data from test set shown on bottom. Length of complete abstract titles: 102.3 characters, 13.34 words (for development set) and 104.0 characters, 13.60 words (for test set)

The performance of the simple NP trimming rule in isolation is shown in Figure 3. Corpus statistics required for the *idf* calculation were extracted from the 10-year MEDLINE collection used in the TREC 2005 genomics track. We varied the *idf* threshold from 2.0 to 10.0 in increments of 0.5, and obtained a plot that relates the BLEU score to the average compression ratio at the character level. For this graph, average compression ratio is computed on all the titles, even those that were unaffected by the rule (since the threshold controls how many titles trigger the rule). The corresponding graph for average compression ratio at the word level looks nearly identical, and is not shown here. It is interesting to note that no threshold actually increases the BLEU score above the baseline (no compression), indicating that users engage in more complex behavior than simply removing modifiers of noun phrases based on *idf* values. Trends observed in the development set carry over to the held-out data, although the rule yields less compression.

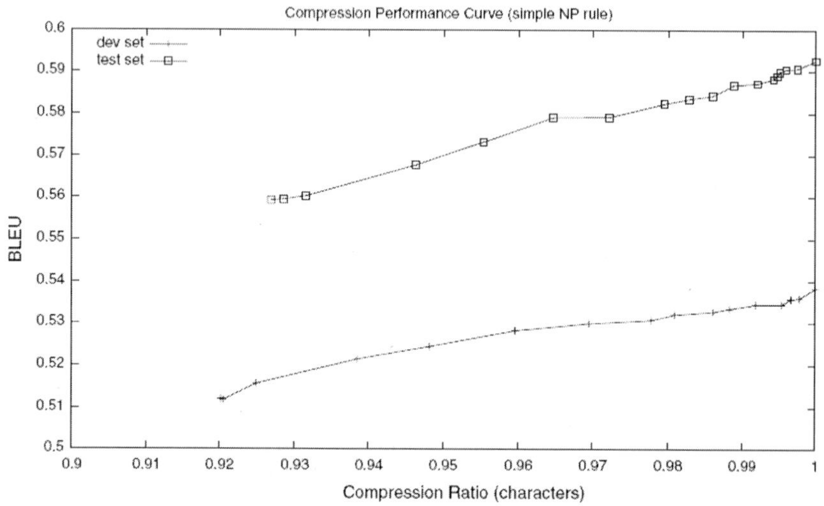

Figure 3. Performance curve of the simple NP rule with *idf* threshold ranging from 2.0 to 10.0 in 0.5 increments

Given that each rule represents a tradeoff between output quality and compression ratio, how can one assemble a complete compression algorithm for MEDLINE abstract titles? In response, we examine sequential application of the trimming rules, which yields curves that characterize the tradeoffs mentioned. In Sect. "Manual evaluation", we report results on manual evaluation of system output at two specific points on this tradeoff curve.

9. SEQUENTIAL APPLICATION OF MULTIPLE RULES

After examining each rule in isolation, we experimented with applying the rules sequentially. The rules were ordered based on BLEU scores on the development set (Table 4): subtitle, DT, VBG, nested PP, serial PP, conjoined NP, and finally simple NP. Results are displayed in Figure 4; the top plot shows compression ratios computed at the character level, and the bottom plot shows compression ratios computed at the word level. The horizontal dotted lines denote the BLEU scores of the original (untrimmed) article titles. The left tails of the curves represent the application of the simple NP rule with different *idf* thresholds (we varied the parameter from 2.0 to 10.0 in increments of 0.5). The labels indicate the points at which each rule is applied (except for the simple NP rule to reduce clutter). In these plots, points closer to the upper left hand corner are "better", in the sense that we desire large reductions in length while maintaining fidelity to the human generated references.

We notice that successive application of individual rules is subjected to a "diminishing returns" effect. The amount of compression achieved with the application of multiple rules is strictly less than the sum of the compression achieved by individual rules. This occurs because earlier rules can eliminate opportunities for later rules to apply. Take the example of prepositional phrases or conjoined NPs inside a gerund phrase. Since the VBG rule applies earlier, the entire phrase would have already been eliminated.

The same general characteristics are observed in both the development and test sets: BLEU scores initially rise and then drop as the rules more and more aggressively trim away parts of the structure. Untrimmed sentences (right edges of the graphs) serve as a baseline—but note that it is possible for our compression algorithm to perform worse if too much is removed from the abstract titles. The dip in the performance curve on the test set can be attributed to the relative performance of the PP trimming rules: one was found to be more effective in the development set, but the reverse turned out to be true in the test set.

How can one interpret these results, especially since BLEU scores do not correspond to any quantity that humans have an intuition for? There are two responses to this question: first, although it is difficult to map absolute scores to a particular level of performance, relative differences in BLEU are meaningful—in that they tell us if one variant is "better" than another (in terms of matching human references).

However, the more appropriate response is to acknowledge the limitations of automatic scoring metrics. Ultimately, our goal is to develop information systems that are useful for humans, and one way of operationalizing "usefulness" is in terms of task performance. Therefore, we believe that the question "How good is a BLEU score of

0.452?" is not pertinent. Rather, we must ask if particular techniques can better assist humans in accomplishing real-world tasks. System development, as guided by automatic metrics, only serves as a stepping stone to extrinsic task-based evaluations. In the next section, we report results from exactly such a study.

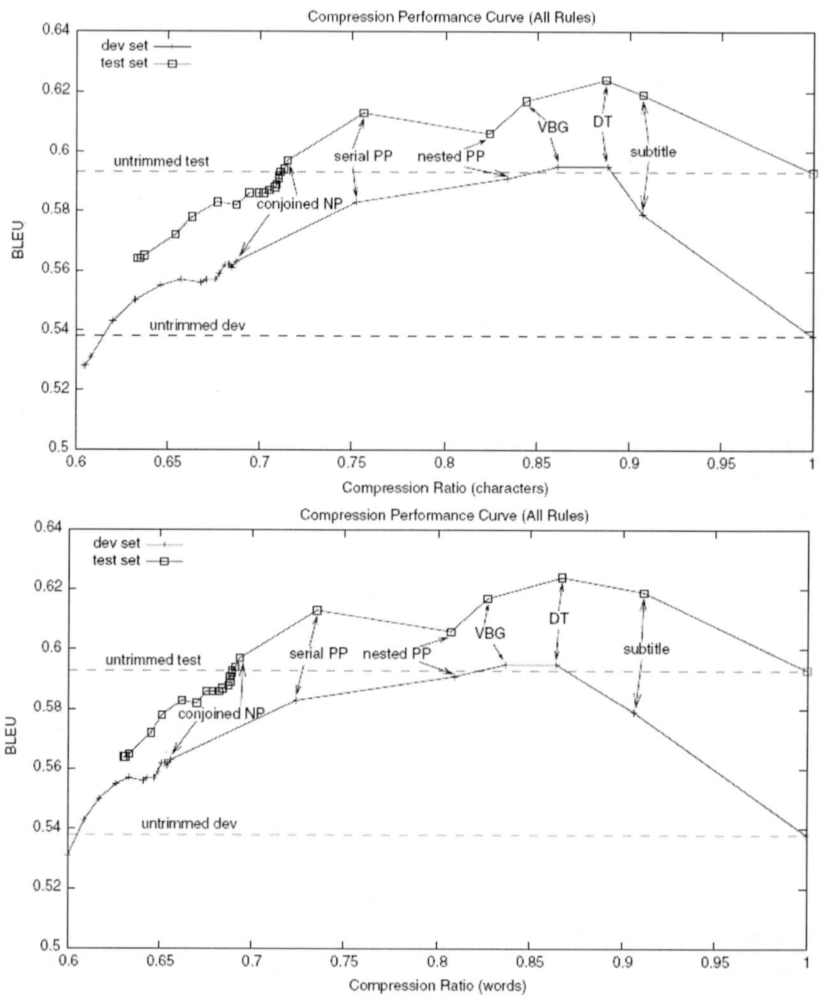

Figure 4. Plots of average compression ratio versus BLEU score, with sequential application of all rules: subtitle, DT, VBG, nested PP, serial PP, conjoined NP, and simple NP (*idf* threshold ranging from 2.0 to 10.0 in 0.5 increments). Average compression ratio in terms of characters shown on top, in terms of words shown on bottom. Horizontal dotted lines show BLEU scores of untrimmed abstract titles

10. MANUAL EVALUATION

In our initial experiments, BLEU primarily served as a formative tool to guide system development. We then conducted a summative task-based evaluation to assess the usefulness of our sentence compression algorithm. At the same time, we also collected human judgments about the intrinsic quality of the compressed output.

Recall from Sect. "Motivation" that we situate sentence compression in the context of related article search in PubMed—in particular, as a method for efficient presentation of items that may be of interest to the user. In this context, the end goal of our sentence compression algorithm is to support interest judgments, that is, a user's decision to examine a citation in detail. The default condition is to show the full title, which serves as a baseline. If the output of our sentence compression algorithm is able to provide the same level of decision support (in terms of accuracy of interest judgments), but with a smaller amount of text, then we can claim to have improved on the baseline. The ability to convey much of the same content in fewer words can be leveraged in two different ways: PubMed can use a smaller screen area for "Related Links", thereby freeing up space or other content elements, or PubMed can display more related articles in the same amount of on-screen space.

Three subject domain experts uninvolved with system development were recruited as assessors. Two of them ("Lo" and "Ly") were the same individuals involved in creating our training and test sets. The third individual ("Wa") was not involved in any other aspect of the project.

Our experiments involved 100 article titles randomly sampled from the development set.5 For each title, we randomly assigned one of four conditions to the trimmed output (25 examples each), described below. The average compression ratio (at the character level) of each condition is shown in the second column of Table 5.

- *Lo:* Manually compressed titles by the annotator "Lo".
- *Ly:* Manually compressed titles by the annotator "Ly".
- *Variant A*: Application of the following rules to the original titles: subtitle, DT, VBG, nested PP, serial PP, and conjoined NP. Note that these titles were longer than the human gold standards on average.

- *Variant B:* Variant A plus the application of the simple NP rule, with an *idf* threshold of 2.0 (aggressive compression, same as the left tail on the plots in Figure 4). On average, these titles were approximately the same length as the human references.

Table 5. Fluency and content judgments by three assessors, on a scale of 1–5 (1 = worst, 5 = best); means and standard deviations are reported

Output	Ratio	Fluency			Content		
		Lo	Ly	Wa	Lo	Ly	Wa
Lo	0.564	3.9 ± 0.95	4.0 ± 1.06	4.0 ± 0.98	3.5 ± 0.96	4.2 ± 0.76	3.6 ± 0.95
Ly	0.575	4.6 ± 0.76	3.0 ± 1.57	4.5 ± 0.87	3.2 ± 1.36	4.3 ± 0.69	3.7 ± 1.06
Variant A	0.687	4.5 ± 0.82	3.4 ± 1.66	4.0 ± 0.93	3.4 ± 1.61	4.0 ± 0.98	3.8 ± 1.18
Variant B	0.563	3.4 ± 1.50	2.8 ± 1.45	3.1 ± 1.36	2.4 ± 1.66	3.5 ± 1.26	3.2 ± 1.19

The second column indicates the mean character compression ratio for each condition

The assessment proceeded in two rounds. In the first round, assessors were provided with the information need and the trimmed title of the abstract. Naturally, they did not know the source of the trimmed titles (i.e., which condition). Assessors were asked to rate the fluency of the title on a scale of 1–5 (1 = worst, 5 = best). They were also asked if they would read the citation in response to the information need (i.e., a judgment of interest).

In the second round, assessors were provided with the information need, the original title, the trimmed title, and their interest judgment from the first round (i.e., response to "Would read abstract?"). They were asked to rate the content of the trimmed title, in terms of capturing essential elements from the full title, on a scale of 1–5 (1 = worst, 5 = best). In addition, they were asked if they would now read the citation given the full title.

This two-phase setup was designed to evaluate both the intrinsic quality of the compressed titles and their effectiveness in a task context. Fluency and content judgments characterize the inherent quality of the compressed titles, while the interest judgments ground our evaluation in a real-world scenario. We were especially concerned about differences in interest judgments from round one and round two. If both responses were the same, we can conclude that the article title was shortened successfully, i.e., the process did not interfere with task performance. A "yes to no" flip provides evidence that the compression process created a false impression of interest. A "no to yes" flip, on the other hand, provides evidence that essential elements from the title were mistakenly removed.6

Fluency and content judgments from our human assessors are provided in Table 5, which shows ratings given by all three assessors on all four conditions (mean and standard deviation). Results suggest that variant A is not any more disfluent than the human-compressed gold standards: two of the three assessors actually placed the machine-generated output ahead of one of the gold standards. Note, however, that variant A titles were on average longer than human references. Machine-generated output of comparable compression (variant B) was found to be consistently less fluent than the other conditions. Similar trends are observed for content ratings: variant A appears to be as good as human output, whereas variant B is less so. Overall, there appears to be much variability in these judgments, suggesting that differences exist in the assessors' interpretation of the task. In particular, we note that Lo preferred Ly's output to Lo's own, in terms of both fluency and content. Similarly, Ly preferred Lo's output to Ly's own (again, both fluency and content). We currently have no reasonable explanation for this observation.

How does the quality of trimmed titles affect users' task performance? The answer can be found in the tally of flips in interest judgments, as shown in Table 6. These results appear to suggest that compressed titles have relatively minimal impact on users' ability to decide whether they want to examine a citation. There does not appear to be much difference between variant A and either one of the human-generated compressions, although variant B titles caused more flips.

Another way of organizing the results is to compare users' interest judgments on the full title (from the second round) with their judgments on compressed titles (from the first round). This is similar to the *consistency test* employed in the TIPSTER SUMMAC evaluations (Mani et al. 2002). Results can be analyzed in terms of the contingency table shown in Table 7, from which we can compute standard aggregate statistics:

$$\text{Precision} = \text{TP}/(\text{TP} + \text{FP})$$
$$\text{Recall} = \text{TP}/(\text{TP} + \text{FN})$$
$$F\text{-score} = 2 \times \text{Precision} \times \text{Recall}/(\text{Precision} + \text{Recall})$$

Table 6. Flips in interest judgments (out of a maximum of 25 for each condition) by three assessors based on compressed (round 1) and full titles (round 2)

Output	Lo		Ly		Wa	
	Y → N	N → Y	Y → N	N → Y	Y → N	N → Y
Lo	2	0	1	1	2	0
Ly	0	1	1	0	0	0
Variant A	0	3	1	0	1	1
Variant B	1	2	2	3	2	2

Table 7. Contingency table for interest judgments on full titles and on compressed variants

Full	Compressed	
	Interesting	¬Interesting
Interesting	True positive (TP)	False negative (FN)
¬Interesting	False positive (FP)	True negative (TN)

Results of this analysis are presented in Table 8. According to the three assessors, variant A titles perform on par with the human-compressed versions, although humans are able to achieve more compression. Variant B titles, which are approximately the same length as the human-generated references, appear to perform worse in terms of precision, recall, and *F*-score. Note that the absolute performance achieved by human-compressed titles and the output of variant A is very high—almost perfect in many cases. This confirms our basic premise that inessential fragments from MEDLINE article titles can be removed without affecting the substance of what is conveyed.

Table 8. P(recision), R(ecall), and *F*(-score) in recognizing interesting citations with four different compressed output, according to three different assessors

Output	Lo			Ly			Wa		
	P	R	*F*	P	R	*F*	P	R	*F*
Lo	0.80	1.00	0.89	0.96	0.96	0.96	0.91	1.00	0.95
Ly	1.00	0.88	0.93	0.93	1.00	0.97	1.00	1.00	1.00
Variant A	1.00	0.73	0.84	0.94	1.00	0.97	0.94	0.94	0.94
Variant B	0.50	0.33	0.40	0.86	0.80	0.83	0.82	0.82	0.82

What are the implications of our findings? Our syntactic compression algorithm is able to shorten abstract titles by approximately 30% without noticeably affecting task performance (variant A). This translates into less material for the user to read, or alternatively, over 40% more content per unit area. In the same space that it takes to display five full abstract titles, we can now display seven. We believe this is a significant result because it provides users access to more potentially interesting articles without requiring them to read more text.

11. DISCUSSION

At a broader level, we believe that this work is significant in two ways. First, the application of syntactic compression techniques in the biomedical domain raises interesting questions about the domain portability of existing language processing tools. Second, we view this work as a case study highlighting the importance of task-based evaluations in grounding summarization technology. This section elaborates on both points.

11.1 Domain adaptation (Or lack thereof)

Issues surrounding the portability of text processing algorithms have recently gained interest in the research community. Due to the availability of corpora and other resources, most modern statistical tools are trained on newswire text, and hence specialized for processing text from that genre, even though many other types of text are worth exploring. Thus, an important consideration in the development of language technologies is its ability to generalize across different domains and genres of text. In this work, we demonstrate that sentence compression techniques originally developed for news articles can be effectively applied to compress MEDLINE article titles. In some ways, this result is somewhat surprising, as we explain below.

First, the biomedical domain offers significant challenges to off-the-shelf parsers. The lexical overlap between MEDLINE abstracts and typical news corpora is surprisingly small (Smith et al. 2005), which creates challenges for parsers—see, for example, (Clegg and Shepherd 2005; Grover et al. 2005; Lease and Charniak2005). As a specific example, prepositional phrase attachment is problematic for statistical parsers trained on newswire text, since noun phrase heads are often unknown lexical items in the biomedical domain. Consider examples taken from Table 2:

- Electroporation-mediated interleukin-12 gene therapy [PP for hepatocellular carcinoma][PP in the mice model].

- Semiquantitative immunoblot analysis [PP of nm23-H1 and -H2 isoforms [PP in adenocarcinomas of the lung]]: prognostic significance.

These abstract titles are typical of those in MEDLINE—characterized by sequences of consecutive prepositional phrases. The examples above are annotated with actual structures assigned by the Stanford Parser, which are correct in both cases. The attachment of PPs, whether to the immediately preceding noun head—the case with example (2)—or another head higher up in the structure—the case with example (1)—is a complex decision that often requires semantic knowledge. Often, the Stanford Parser is incorrect in its choice of PP attachment point, as illustrated by the following parse:

- Induction [PP of cell cycle arrest and morphological differentiation [PP by Nurr1 and retinoids]] [PP in dopamine MN9D cells].

Rather, the correct structure should be:

- Induction [PP of cell cycle arrest and morphological differentiation][PP by Nurr1 and retinoids][PP in dopamine MN9D cells].

Nevertheless, our nested and serial PP rules appear to be insensitive to these errors because they were engineered by examining Stanford Parser output. Since the parser appears to make systematic errors, we are still able to capture generalizations—albeit these rules may not represent linguistically valid generalizations in the biomedical domain. In example (3), the serial PP rule removes the PP "in dopamine MN9D cells", which appears to yield a reasonable compression.

Second, MEDLINE article titles are quite different from sentences that occur in newswire text—most of the time, titles are not even complete sentences. Since titles are often noun phrases or verb phrases, our approach must not only cope with out-of-domain effects (most notably, unknown lexical items), but also stylistic differences. Experiments suggest that the syntactic trimming approach is also capable of handling such divergences, given a set of rules developed specifically for the biomedical domain. It is noteworthy that respectable performance is achieved in our application without any domain adaptation.

11. 2 Grounding summarization in real-world tasks

This work serves as a case study illustrating the importance of grounding summarization tasks in real-world user scenarios. To a human, a fluency score of four (out of five) is not particularly meaningful, and neither is a 0.314 BLEU score. However, quantifying performance in terms of decision-making accuracy on compressed titles (as compared to the full titles) is informative because it illustrates how summarization techniques assist the user's end task, that of knowledge exploration and information gathering. In general, we believe that information presentation issues provide a general framework for task-based evaluation of summarization systems; see also, (Mani et al. 2002; Dorr et al. 2005) for similar setups.

We would like to end this section with a discussion of our underlying task model. In most retrieval tasks, the assumption is that the user issues a query to a search engine and obtains a ranked list of documents that are potentially relevant. This output then serves as the starting point to browsing, selection, examination, and query reformulation mechanisms that may ultimately lead to the satisfaction of the information need. However, this traditional query-centered model does not describe the only possible pattern of user-system interactions. In particular, PubMed attempts to draw connections between articles in

MEDLINE by unobtrusively displaying titles that may be of interest. This mechanism provides users with another device for exploring the information space. In fact, previous simulation studies have shown that such a feature can improve performance, as measured by traditional ranked retrieval metrics (Wilbur and Coffee 1994; Smucker and Allan2006). Although we focus primarily on the related links features in PubMed, our syntactic compression techniques are equally applicable to other components in the retrieval environment, e.g., summarizing ranked lists so that more results can be displayed on any given Web page.

12. FUTURE WORK

With respect to syntactic compression in the biomedical domain, there are two distinct threads of future work worth exploring—improvements to the compression algorithm and application of similar techniques to related problems. We briefly discuss each.

The simple NP rule is currently the only rule that is parameterized—in this case, an*idf* threshold. The same idea could be applied to other rules.7 For example, the system could use a threshold to determine if the head of a prepositional phrase was "lightweight", and then factor in this evidence to determine if the PP was removable. In the same way, the conjoined NP rule could also be subjected to this modification.

However, the introduction of parameterized rules adds additional complexity to rule ordering. Currently, trimming rules are applied sequentially in a fixed order (based on individual performance in isolation). This neglects possible interaction effects between rules, which would certainly increase with the introduction of parameters. Zajic (2007) explored a solution to a similar problem by allowing multiple simultaneous rule applications, and then developing a mechanism to select among the multiple compressed candidates. We believe that the same idea can be applied here.

Beyond applications in information retrieval, sentence compression techniques can also be used for other tasks in the biomedical domain. The extraction of GeneRIFs is one such possibility.8 GeneRIFs are concise phrases describing a function of a gene, explicitly linked to the Entrez Gene database. The extraction and linking of these descriptions is important for biologists and other researchers, since such information would otherwise be scattered in many disparate articles and sources. GeneRIFs by design are limited to 255 characters,9 so brevity is highly desired. As previously discussed, summarization techniques have been successfully applied to extracting GeneRIFs (Ling et al. 2006; Lu et al. 2006)—these systems could additionally benefit from the syntactic compression techniques discussed in this article, to eliminate inessential material from the extracted phrases.

13. CONCLUSION

Like much previous work, this paper examines the task of sentence compression. However, our perspective is novel in two different ways: we explore the problem in the domain of biomedicine and within the context of an information retrieval task. The contributions of this study are two-fold: first, we demonstrate that the syntactic trimming approach, which has proven effective in the newswire domain, is portable to the biomedical domain. Second, our work highlights the importance of extrinsic evaluations and grounding summarization in real-world tasks. It is our hope that we can eventually transition summarization technology into the PubMed search engine.

FOOTNOTES

[1]http://www.ncbi.nlm.nih.gov/entrez/

[2]http://www.nlm.nih.gov/pubs/factsheets/medline.html

[3]Version 1.5.1, downloaded from http://www-nlp.stanford.edu/software/; default settings were used in all experiments.

[4]On a sentence-by-sentence basis, that is. Discourse-level properties such as coherence of text are important issues, but much harder to evaluate and quantify.

[5]We avoided using the held-out test data so that we can continue developing the trimming algorithm in the future. This does not impact our findings, since the development-test division is meaningless from our assessors' point of view.

[6]Note that since the evaluated set was constructed with a balanced number of relevant and irrelevant articles (as deterimined by the original genomics track assessors), the consistency measures are not distorted by class imbalance issues.

[7]We would like to thank an anonymous reviewer for pointing this out.

[8]We would again like to thank an anonymous reviewer for pointing this out.

[9]http://www.ncbi.nlm.nih.gov/projects/GeneRIF/GeneRIFhelp.html

ACKNOWLEDGMENTS

We would like to thank our annotators and assessors for their efforts, without which this study would not be possible. Our thanks also go out to the anonymous reviewers, whose comments and criticisms have made this article significantly better. For this work, JL was funded in part by the National Library of Medicine, where he was a visiting research scientist during the summer of 2006. WJW is supported by the Intramural Research Program of the NIH, National Library of Medicine. JL would also like to thank Esther and Kiri for their kind support.

REFERENCES

1. Afantenos, Stergos, Karkaletsis, Vangelis, & Stamatopoulos, Panagiotis (2005). Summarization from medical documents: A survey. *Artificial Intelligence in Medicine, 33*(2), 157–177.

2. Aronson, Alan R., Mork, James G., Gay, Clifford W., Humphrey, Susanne M., & Rogers, Willie J. (2004). The NLM indexing initiative's medical text indexer. In *Proceedings of the 11th World Congress on Medical Informatics (MEDINFO 2004)* (pp. 268–272). San Francisco, California.

3. Banko, Michele, Mittal, Vibhu, & Witbrock, Michael (2000). Headline generation based on statistical translation. In*Proceedings of the 38th*

Annual Meeting of the Association for Computational Linguistics (ACL 2000) (pp. 318–325). Hong Kong, China.

4. Bergler, Sabine, Witte, René, Khalife, Michelle, Li, Zhuoyan, & Rudzicz, Frank (2003). Using knowledge-poor coreference resolution for text summarization. In *Proceedings of the HLT/NAACL 2003 Text Summarization Workshop and Document Understanding Conference (DUC 2003)* (pp. 85–92). Edmonton, Alberta.

5. Blair-Goldensohn, Sasha, Evans, David, Hatzivassiloglou, Vasileios, McKeown, Kathleen, Nenkova, Ani, Passonneau, Rebecca, Schiffman, Barry, Schlaikjer, Andrew, Siddharthan, Advaith, & Siegelman, Sergey (2004). Columbia University at DUC 2004. In *Proceedings of the 2004 Document Understanding Conference (DUC 2004) at HLT/NAACL 2004* (pp. 23–30). Boston, Massachusetts.

6. Brandow, Ronald, Mitze, Karl, & Rau, Lisa F. (1995). Automatic condensation of electronic publications by sentence selection. *Information Processing and Management, 31*(5), 675–685.

7. Clegg, Andrew B., & Shepherd, Adrian (2005). Evaluating and integrating Treebank parsers on a biomedical corpus. In *Proceedings of the ACL 2005 Workshop on Software*. Ann Arbor, Michigan.

8. Conroy, John M., Schlesinger, Judith D., & Stewart, Jade Goldstein (2005). CLASSY query-based multi-document summarization. In *Proceedings of the 2005 Document Understanding Conference (DUC-2005) at NLT/EMNLP 2005*. Vancouver, Canada.

9. Dorr, Bonnie J., Zajic, David, & Schwartz, Richard (2003). Hedge trimmer: A parse-and-trim approach to headline generation. In *Proceedings of the HLT/NAACL 2003 Text Summarization Workshop and Document Understanding Conference (DUC 2003)* (pp. 1–8). Edmonton, Alberta.

10. Dorr, Bonnie J., Monz, Christof, President, Stacy, Schwartz, Richard, & Zajic, David (2005). A methodology for extrinsic evaluation of text summarization: Does ROUGE correlate? In *Proceedings of the ACL*

2005 Workshop on Intrinsic and Extrinsic Evaluation Measures for MT and/or Summarization. Ann Arbor, Michigan.

11. Elhadad, Noemie, Kan, Min-Yen, Klavans, Judith, & McKeown, Kathleen (2005). Customization in a unified framework for summarizing medical literature. *Journal of Artificial Intelligence in Medicine, 33*(2), 179–198.

12. Grover, Claire, Lascarides, Alex, & Lapta, Mirella (2005). A comparison of parsing technologies for the biomedical domain. *Natural Language Engineering, 11*(1), 27–65.

13. Hersh, William, Cohen, Aaron, Yang, Jianji, Bhupatiraju, Ravi, Roberts, Phoebe, & Hearst, Marti (2005). TREC 2005 genomics track overview. In *Proceedings of the Fourteenth Text REtrieval Conference (TREC 2005)*. Gaithersburg, Maryland.

14. Jing, Hongyan (2000). Sentence reduction for automatic text summarization. In *Proceedings of the 6th Applied Natural Language Processing Conference (ANLP'00)*. Seattle, Washington.

15. Kan, Min-Yen, McKeown, Kathleen R., & Klavans, Judith L. (2001). Domain-specific informative and indicative summarization for information retrieval. In *Proceedings of the 2001 Document Understanding Conference (DUC 2001) at SIGIR 2001*. New Orleans, Louisiana.

16. Klein, Dan, & Manning, Christopher D. (2003). Accurate unlexicalized parsing. In *Proceedings of the 41st Annual Meeting of the Association for Computational Linguistics (ACL 2003)* (pp. 423–430). Sapporo, Japan.

17. Knight, Kevin, & Marcu, Daniel (2000). Statistics-based summarization—step one: Sentence compression. In*Proceedings of the Seventeenth National Conference on Artificial Intelligence (AAAI-2000)* (pp. 703–710). Austin, Texas.

18. Kupiec, Julian, Pedersen, Jan O., & Chen, Francine (1995). A trainable document summarizer. In *Proceedings of the 18th Annual*

International ACM SIGIR Conference on Research and Development in Information Retrieval (SIGIR 1995) (pp. 68–73). Seattle, Washington.

19. Lease, Matthew, & Charniak, Eugene (2005). Parsing biomedical literature. In *Proceedings of the Second International Joint Conference on Natural Language Processing (IJCNLP'05)* (pp. 58–69). Jeju Island, Korea.

20. Lin, Chin-Yew, & Hovy, Eduard (2003). Automatic evaluation of summaries using *n*-gram co-occurrence statistics. In *Proceedings of the 2003 Human Language Technology Conference and the North American Chapter of the Association for Computational Linguistics Annual Meeting (HLT/NAACL 2003)* (pp. 71–78). Edmonton, Alberta.

21. Lin, Chin-Yew (2004). ROUGE: A package for automatic evaluation of summaries. In *Proceedings of the Workshop on Text Summarization Branches Out (WAS 2004) at ACL 2004*. Barcelona, Spain.

22. Ling, Xu, Jiang, Jing, He, Xin, Mei, Qiaozhu, Zhai, Chengxiang, & Schatz, Bruce (2006). Automatically generating gene summaries from biomedical literature. In *Pacific Symposium on Biocomputing, 11*, 40–51.

23. Lu, Zhiyong, Cohen K. Bretonnel, & Hunter, Lawrence (2006). Finding GeneRIFs via gene ontology annotations. In*Pacific Symposium on Biocomputing 11*, pp. 52–63.

24. Mani, Inderjeet, Gates, Barbara, & Bloedorn, Eric (1999). Improving summaries by revising them. In *Proceedings of the 37th Annual Meeting of the Association for Computational Linguistics (ACL 1999)* (pp. 558–565). College Park, Maryland.

25. Mani, Inderjeet, Klein, Gary, House, David, & Hirschman, Lynette (2002) SUMMAC: A text summarization evaluation. *Natural Language Engineering, 8*(1), 43–68.

26. McKeown, Kathleen R., Elhadad, Noemie, & Hatzivassiloglou, Vasileios (2003). Leveraging a common representation for personalized search and summarization in a medical digital library.

In *Proceedings of the 3rd ACM/IEEE-CS Joint Conference on Digital Libraries (JCDL 2003)*. (pp. 159–170). Houston, Texas.

27. Minel, Jean-Luc, Nugier, Sylvaine, & Piat, Gerald (1997). How to appreciate the quality of automatic text summarization? Examples of FAN and MLUCE protocols and their results on SERAPHIN. In *Proceedings of the ACL/EACL 1997 Workshop on Intelligent Scalable Text Summarization*. Madrid, Spain.

28. Mizzaro, Stefano (1998). How many relevances in information retrieval? *Interacting With Computers, 10*(3), 305–322.

29. Morris, Andrew, Kasper, George, & Adams, Dennis (1992). The effects and limitations of automated text condensing on reading comprehension performance. *Information Systems Research, 3*(1), 17–35.

30. Paice, Chris D. (1990). Constructing literature abstracts by computer: Techniques and prospects. *Information Processing and Management, 26*(1), 171–186.

31. Papineni, Kishore, Roukos, Salim, Ward, Todd, & Zhu, Wei-Jing (2002). BLEU: A method for automatic evaluation of machine translation. In *Proceedings of the 40th Annual Meeting of the Association for Computational Linguistics (ACL 2002)* (pp. 311–318). Philadelphia, Pennsylvania.

32. Ruch, Patrick, Chichester, Christine, Cohen, Gilles, Coray, Giovanni, Ehrler, Frédéric, Ghorbel, Hatem, Müller, Henning, & Pallotta, Vincenzo (2003). Report on the TREC 2003 experiment: Genomic track. In *Proceedings of the Twelfth Text REtrieval Conference (TREC 2003)*. Gaithersburg, Maryland.

33. Salton, Gerard, Singhal, Amit, Mitra, Mandar, & Buckley, Chris (1997). Automatic text structuring and summarization. *Information Processing and Management, 33*(2), 193–207.

34. Smith, Lawrence H., Rindflesch, Thomas C., & Wilbur, W. John. (2005). The importance of the lexicon in tagging biological text. *Natural Language Engineering, 12*(2), 1–17.

35. Smucker, Mark, & Allan, James (2006). Find-Similar: Similarity browsing as a search tool. In *Proceedings of the 29th Annual International ACM SIGIR Conference on Research and Development in Information Retrieval (SIGIR 2006)* (pp. 461–468). Seattle: Washington.

36. Turner, Jenine, & Charniak, Eugene (2005). Supervised and unsupervised learning for sentence compression. In*Proceedings of the 43rd Annual Meeting of the Association for Computational Linguistics (ACL 2005)* (pp. 290–297). Ann Arbor, Michigan.

37. Wang, Ruichao, Stokes, Nicola, Doran, William, Newman, Eamonn, Carthy, Joe, & Dunnion, John (2005). Comparing Topiary-style approaches to headline generation. In *Lecture Notes in Computer Science: Advances in Information Retrieval: 27th European Conference on IR Research (ECIR 2005)*, Vol. 3408. Santiago de Compostela, Spain: Springer Berlin/Heidelberg.

38. Wilbur, W. John, & Coffee, Leona (1994). The effectiveness of document neighboring in search enhancement.*Information Processing and Management, 30*(2), 253–266.

39. Wilbur, W. John (2005). Modeling text retrieval in biomedicine. In Chen, Hsinchun, Fuller, Sherrilynne S., Friedman, Carol, & Hersh, William (Eds.), *Medical informatics: Knowledge management and data mining in biomedicine* (pp. 277–297). New York: Springer Science.

40. Zajic, David, Dorr, Bonnie J., & Schwartz, Richard (2004). BBN/UMD at DUC-2004: Topiary. In *Proceedings of the 2004 Document Understanding Conference (DUC 2004) at NLT/NAACL 2004*. Boston, Massachusetts.

41. Zajic, David, Dorr, Bonnie, Lin, Jimmy, & Schwartz, Richard (2007). Multi-candidate reduction: Sentence compression as a tool for

document summarization tasks. *Information Processing and Management, Special Issue on Summarization*, in press.

42. Zajic, David M. (2007). *Multiple Alternative Sentence Compressions (MASC) as a tool for automatic summarization tasks*. Ph.D. thesis, College Park: University of Maryland.

43. Zhou, Liang, & Hovy, Eduard (2003). Headline summarization at ISI. In *Proceedings of the HLT-NAACL 2003 Text Summarization Workshop and Document Understanding Conference (DUC 2003)* (pp. 174–178). Edmonton, Alberta.

2

EVALUATION OF QUERY EXPANSION USING MESH IN PUBMED

Zhiyong Lu , Won Kim, W. John Wilbur

National Center for Biotechnology Information (NCBI),
National Library of Medicine, Bethesda, MD 20894, USA

ABSTRACT

This paper investigates the effectiveness of using MeSH® in PubMed through its automatic query expansion process: Automatic Term Mapping (ATM). We run Boolean searches based on a collection of 55 topics and about 160,000 MEDLINE®citations used in the 2006 and 2007 TREC Genomics Tracks. For each topic, we first automatically construct a query by selecting keywords from the question. Next, each query is expanded by ATM, which assigns different search tags to terms in the query. Three search tags: [MeSH Terms], [Text Words], and [All Fields] are chosen to be studied after expansion because they all make use of the MeSH field of indexed MEDLINE citations. Furthermore, we characterize the two different mechanisms by which the MeSH field is used. Retrieval results using MeSH after expansion are compared to those solely based on the words in MEDLINE title and abstracts. The aggregate retrieval performance is assessed using both F-measure and mean rank precision. Experimental results suggest that query expansion using MeSH in PubMed can generally improve retrieval performance, but the improvement may not affect end PubMed users in realistic situations.

KEYWORDS

Query expansion; TREC genomics; PubMed; MeSH

1 INTRODUCTION

Query expansion refers to the process of reformulating a query to improve retrieval performance, typically by including additional terminology that is synonymous. PubMed, the search engine for over 17 million biomedical and health citations in MEDLINE, allows its users to manually attach search field tags (also called qualifiers) to search terms in square brackets (i.e., search term [tag]) in order to improve on users' original searches. However, such advanced MEDLINE search features were seldom used as seen in the PubMed log analysis (Herskovic et al.2007). Alternatively, PubMed employs a process called Automatic Term Mapping (ATM) that compares and maps untagged terms from the user query to lists of pre-indexed terms in PubMed's translation tables (in this order): the Medical Subject Headings (MeSH) table (mapping search terms to MeSH concepts), the journals translation table (mapping search terms to journal names), and the author index (mapping search terms to author names). In the context of query expansion, we consider only the translation via MeSH here, which is designed by the National Library of Medicine (NLM) for indexing and searching of the MEDLINE database of journal citations. That is, if a user query includes a term that can be mapped to a MeSH concept, the search will then add the MeSH term to the original query. As a result, the ATM process enables the original query to have an access to the MeSH field of indexed MEDLINE documents. For instance, if a user submits a query*tumor* with no search tags attached, the query will be automatically mapped to the MeSH term NeoplasmsNeoplasms by ATM. Thus, not only documents having the word*tumor* in the title and abstract will be retrieved, any other documents indexed with the MeSH term NeoplasmsNeoplasms (as well as more specific terms beneath it in the MeSH hierarchy) will be returned as well. Furthermore, individual words in MeSH concepts are also indexed and searchable. Take the previous query for example, documents indexed with MeSH terms that include the word *tumor* (e.g. Tumor Virus InfectionsTumor Virus Infections) will also be returned.

By default, PubMed applies the Automatic Term Mapping feature to every unqualified user query, thus affecting the search results of millions of user queries each day in spite of the fact that many users may not actually realize its existence. It is designed and implemented by the NLM as a means to improve retrieval performance. Unlike the studies in the past that focused on the accuracy of the mapping process (Gault et al. 2002; Carlin 2004; Smith 2004; Shultz 2006), the main objective of this work is to examine the effectiveness of using the MeSH field of indexed documents for improving retrieval performance.

1.1 Related work

There is a significant body of literature on query expansion as an effective technique in the general information retrieval (IR) community (Salton and Buckley1997; Mitra et al. 1998). However, in the domain of biomedical text retrieval, results have been mixed. Earlier research has suggested the use of this technique could result in elevated retrieval performance (Srinivasan 1996; Aronson and Rindflesch 1997), while no performance gain was seen in others (Hersh et al.2000).

More recently, this technique has been widely implemented in various retrieval systems built by a number of participating teams in TREC Genomics tracks since 2003 (Hersh and Bhupatiraju 2003). Like previous investigations, there have been some contradictory reports on the benefits of using query expansion. Results from some TREC Genomics *ad hoc* retrieval task participants (Hersh and Bhupatiraju2003; Abdou et al. 2005) pointed to the use of query expansion through multiple concepts: MeSH headings, substance name fields in MEDLINE, and species as accounting for improved performance. On the contrary, other participating teams (Hersh et al. 2003, 2004, Guo et al. 2004; Bacchin and Melucci 2005) reported no or detrimental impact when expanding original queries with multiple sources including MeSH concepts.

The main differences between the work presented in this paper and the previous studies are (1) the retrieval system being used during evaluation, and (2) the method used for query expansion. All previous analyses based their results on either the vector space model (Salton and Buckley 1988) or probabilistic models like the Okapi model (Robertson and Walker 1994), both of which rank retrieved documents on the basis of their relevance to the queries. In contrast, PubMed is a Boolean search system and displays retrievals in reverse chronological order by default. Thus, we implemented our own search system that supports both Boolean and probabilistic text retrievals. As a result, not only can we evaluate the query expansion approach in the same setting as used by PubMed, but we can compare retrieval results with term weighting ranking strategies demonstrated in the previous studies.

The second distinction lies in the methods used for query expansion. The expansion terms in general come from two different sources. First, queries can be expanded using thesauri like the Unified Medical Language System (UMLS) Metathesaurus (Lindberg et al. 1993) via either computational approaches (Srinivasan 1996; Aronson 1996; Guo et al. 2004; Abdou et al. 2005) or manual mappings (Hersh et al. 2000). Second, expansion terms can be extracted using retrieval feedback (Srinivasan 1996; Bacchin and Melucci 2005). In our investigation, query expansion was executed through ATM, a unique mapping process of PubMed.

Finally, to the best of our knowledge, this is the first formal evaluation on the benefits of applying the query expansion technique to a daily operational search system as opposed to retrieval systems mainly designed and tested in research laboratories. Therefore, our analysis plays a critical role in the understanding of future technology development needs for PubMed, along with its mission to better fulfill the information needs of millions of PubMed users.

2 METHODS

2.1 Text collection

In this work, we used the TREC Genomics data in both 2006 and 2007 (Hersh et al. 2006, 2007), which consist of 28 topics and 36 topics (in the form of biological questions), respectively. In both years, a total of 162,048 full-text documents from Highwire Press1 were used. Except for 1,800 instances, most of the documents were successfully mapped to their corresponding PubMed Identifiers (PMIDs). Hereafter, we refer to the remaining set of 160,248 PMIDs as the *TREC set* in our study. The 36 topics (Topic IDs 200 to 235) in 2007 and 28 (Topic IDs 160 to 187) in 2006 were collected from bench biologists and represent actual information needs in biomedical research. For demonstration purposes, we show three topics 207, 229 and 231 from TREC 2007 in the list below:

<207> What toxicities are associated with etidronate?

<229> What signs or symptoms are caused by human parvovirus infection?

<231> What tumor types are found in zebrafish?

For each topic, a set of relevant documents from the TREC set were produced by the relevance judgments based on pooled results from team submissions. They are assumed as the ground truth or gold standard data in our investigation and referred to as the *relevant set* in the remainder of this paper.

2.2 Query construction

For each TREC topic, a user query is required for retrieving relevant documents in PubMed. In order to produce unbiased (i.e. without human interference) user queries in a consistent manner, we chose to automatically select words from questions as queries on the assumption that real users would also intuitively create their queries based on the questions. Specifically, for each question, we first removed stop words (Wilbur and Sirotkin 1991) from the question and enumerated all possible word combinations, each of which was then expanded by ATM and subsequently searched in PubMed. Next, for each word combination that retrieved a non-empty set of documents, we compared those retrieved documents to the ones in the relevant set and computed the standard IR measures (Hersh 2003): recall, precision and F-measure (see also: Sect. 3.1). In the end, the query with the highest F-measure was selected.

Take the topic 229 for example, we first removed stop words *what, or, are, by* from the question. A total of 63 different user queries were then generated based on the remaining six words (*signs, symptoms, caused, human, parvovirus, infection*) and subsequently searched in PubMed to obtain a set of relevant documents. Three example queries are shown in Table 1, together with their corresponding IR measures after comparing the retrieved set with the relevant set. The query *human parvovirus* was finally chosen for further study because it yielded the highest F-measure. We processed all 64 topics in this way and were able to identify query terms for most of the topics except 9 instances (topics 164, 173, 177, 179, 180, 184, 185 in 2006 and topics 207 and 225 in 2007) where either no relevant documents were found in the gold standard or no query terms could be generated to represent meaningful user queries (i.e. using query expansion their F-measures are almost zero). Therefore, these nine topics were excluded from further analysis. The automatically generated queries for the remaining 55 TREC topics varied in length from a single word to a maximum of four words, with a mean of 2.5 words per query.

Table 1. Three potential user queries built automatically by selecting words from topic 229

Word combination	Recall	Precision	F-measure
human parvovirus	0.47	0.63	0.54
human parvovirus infection	0.39	0.85	0.53
parvovirus infection	0.39	0.69	0.49

2.3 Preprocessing search field tags after translation

Once all of the 55 user queries were determined, their translated counterparts through PubMed's Automatic Term Mapping were obtained. For instance, *human parvovirus* was translated to: ((“humans”[TIAB] NOT Medline[SB]) OR “humans”[MeSH Terms] OR human[Text Word]) AND (“parvovirus”[MeSH Terms] OR parvovirus[Text Word]).2 Translated queries may contain a variety of different search field tags. We filtered all but three tags: [MeSH Terms], [Text Words] and [All Fields] because they are (a) the most frequent tags seen after expansion (see also: Sect. 4.1); and (b) the major ones involving the use of MeSH. Therefore, they are the only important ones in our analysis for comparing results with respect to query expansion. For example, the function of the Subset search field ([SB]) is to restrict retrieval by topic, citation status and journal/citation subsets. Thus, the “human"[TIAB] NOT Medline[SB]” part of the translation for *human parvovirus*refers to searches for articles in PubMed but not yet indexed for MEDLINE (e.g. in-process citations which have been provided a PMID before it is indexed with MeSH). Since all of the citations in TREC 2006 and 2007 have already been indexed at the time of our experiments, this specific search tag is not applicable to our data. Thus, it was discarded. As a result, the remaining part: “humans"[MeSH

Terms] OR human[Text Word]) AND ("parvovirus"[MeSH Terms] OR parvovirus[Text Word]) served as the input to retrieve citations in our system.

All three search tags make use of the MeSH fields. However, the way they use the MeSH field varies. Based on the online help documents3 and discussion with an expert medical librarian at the National Library of Medicine (Kathi Canese, personal communication), we treated the three search tags as follows in this work:

1) [Text Words]: Match against all words in MEDLINE title, abstract, and MeSH terms (no automatic MeSH explosion).

2) [MeSH Terms]: Match against the MeSH term and the more specific ones beneath it in the MeSH hierarchy (i.e. automated MeSH explosion).

3) [All Fields]: Same as [Text Words].

For the search tags [Text Words] and [MeSH Terms], they were handled identically in our analysis as they would be in PubMed. For [All Fields], we limited search fields to text words only because other bibliographic information of a MEDLINE citation (e.g. author name) was not used by our search system. A word or phrase will only be tagged with [All Fields] if no match can be found in existing translation tables. For instance, if a user query is *lysosomal*, it will be translated to lysosomal [All Fields] since no match can be found for it in any of the translation tables in PubMed.

As can be seen, PubMed would only retrieve documents matching the search keywords in the title or abstracts if the ATM was not used. After applying ATM, PubMed expands its search ability by examining the MeSH field of indexed documents, which could result in returning many relevant documents that would otherwise be missed. For each retrieved document in the following analyses, we were able to characterize whether it was retrieved because of the use of MeSH or because of word match in the title or abstract. In addition, for the ones that were retrieved because of MeSH, we further classified them into three different groups: matching a MeSH term, matching a more specific MeSH term due to automatic MeSH explosion, or matching words of a MeSH term (See also: Sect. 4.2). In the following experiments, query expansion refers to the inclusion of all three types of MeSH related retrievals.

3 RESULTS

3.1 Evaluation by recall, precision and F-measure

Many different measures for evaluating the performance of IR systems have been proposed and used in evaluating the effectiveness of query expansion (Hersh2003), two of which are selected in this study: *F-measure* and *mean rank precision*.

Retrieval performance was first assessed by recall (the proportion of relevant documents in the collection retrieved by the query) and precision (the proportion of retrieved documents relevant to the query). Then, F-measure was computed as the harmonic mean of recall and precision as follows:

$$F - measure = \frac{2 \times (precision \times recall)}{precision + recall} \tag{1}$$

As can be seen in Table 2, results using query expansion are better than those without the query expansion in terms of average F-measures in both TREC 2006 and 2007. Table 3 serves as a query-by-query F-measure analysis, describing the number of queries for which the query expansion technique achieved better, worse, or equal performance levels as done without the technique. The last column in Table 3 lists the results of the non-parametric statistical tests (Noreen 1989; Wilbur1994) we performed through the pair-wise comparisons of all of the individual F-measures. Discrepancies between the results of 2006 and 2007 may be mostly attributed to the differences in topics and possibly the quality of relevance judgments. Our analysis shows that the relevant sets of TREC 2006 and 2007 consist of 1,448 and 2,478 documents, respectively. In contrast, about the same number of total relevant documents were retrieved (e.g., 661 in 2006 and 662 in 2007) when query expansion was used. As a result, performance (as shown in F-measures in Table 2) is considerably higher in 2006 than in 2007.

As the statistical test does not account for individual differences in magnitude, Figure 1 and 2 show that for topics in TREC 2006 and 2007, the enhancements in F-measure are generally far greater than are the degradations. The largest performance improvement was observed to be over 7 fold. Without using query expansion, the F-measure for Topic 200 (see Table 4) in TREC 2007 was 0.047. With query expansion, the F-measure increased substantially to 0.398.

Table 2. Results of average F-measures on the 2006 and 2007 TREC topics

TREC	Topics	Query expansion	Average F-measures
2006	21	Yes	0.406
		No	0.334
2007	34	Yes	0.264
		No	0.214

Table 3. Query-by-query F-measure analysis: with versus without query expansion (confidence level = 0.05)

TREC	Better	Worse	Equal	Statistical significance
2006	7	5	9	No
2007	20	8	6	Yes

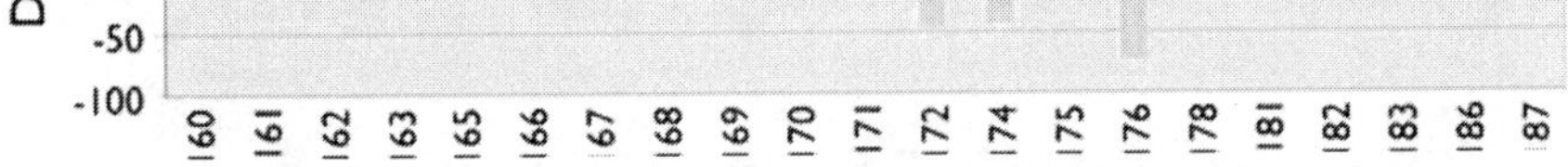

Figure 1. TREC 2006: Query-by-query differences in F-measures

While we can see performance enhancement for most of the topics, query expansion resulted in degraded performance for some topics. The largest performance decrease happened to Topic 176 (see Table 4) where its F-measure dropped from 0.046 to 0.017 (a 72.6% decrease) after applying the query expansion technique. Further analysis showed that the total number of retrieved documents increased over 10 fold (from 21 to 221) due to an expansion of the term: *degradation*. PubMed translated the text word *degradation* to "metabolism" [Subheading] OR ("metabolism" [TIAB] NOT Medline[SB]) OR "metabolism" [MeSH Terms] OR degradation[Text Word]. For this topic, this expansion was harmful to the performance.

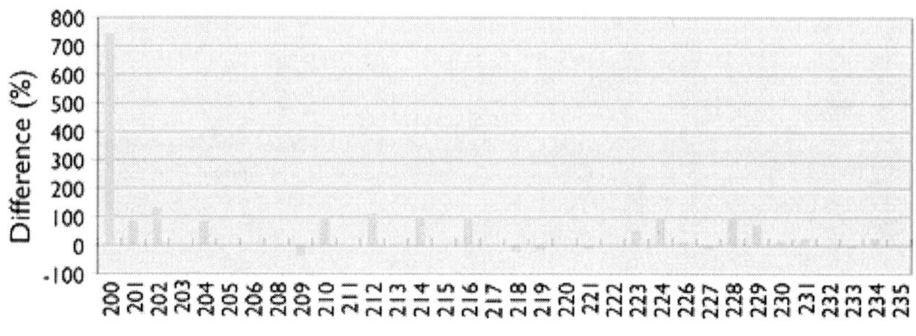

Figure 2. TREC 2007: Query-by-query differences in F-measures

Table 4. The two topics (176 and 200) that display the largest performance improvement and degradation after applying the query expansion technique

Topic ID	Question	Constructed query
176	How does Sec61-mediated CFTR degradation contribute to cystic fibrosis?	CFTR degradation cystic fibrosis
200	What serum proteins change expression in association with high disease activity in lupus	Proteins lupus

3.2 Evaluation by mean rank precision

Our main concern in practice is to enhance retrieval performances among the top few returned documents which users are most likely to examine. To analyze this aspect, we introduce the second measurement for our experiments: Mean Rank Precision (MRP), which is the mean value of the rank precisions (precision at a given rank) computed for all queries. In this study, we chose the cut-off ranks to be 5, 10, and 20 as most of the retrievals occurred before 20 based on our own experience with PubMed users. Thus, the mean rank precision reveals more directly how a ranking strategy affects user retrieval effectiveness in realistic situations.

Note that PubMed returns citations in reverse chronological order by default, displaying more recent citations earlier. In accordance with this display order, we first sorted PMIDs numerically and then reversed the order. As a result, larger PMIDs would appear earlier in a ranked list.

Results in Table 5 show the mean average precisions at given ranks (5, 10, and 20) sorted by the reverse time order, suggesting the existing query expansion feature in PubMed does not always yield improved performance. In fact, the difference between results with and without query expansion is as small as a single relevant document for the TREC 2006 data. Additionally, results from TREC 2007 demonstrate that using such a technique could sometimes be harmful to system performance. This can happen because query expansion is generally considered as a recall-favoring technique (as opposed to precision-favoring) that often results in a large number of additional retrievals. For instance, the number of retrieved documents increased from 1,174 to 3,163 when the query expansion technique was employed in TREC 2007. A much smaller increase in total number of retrievals (from 1,463 to 1,899) was seen in TREC 2006. This may account for the differences between 2006 and 2007, as shown in Table 5.

Table 5. Results of the 2006 and 2007 TREC topics with and without query expansion as measured by mean rank precision

TREC	Query expansion	P@5	P@10	P@20
2006	Yes	0.556	0.540	0.540
	No	0.523	0.517	0.524
2007	Yes	0.385	0.411	0.413
	No	0.434	0.454	0.452

Retrieved documents are sorted in reverse chronological order

3.3 Results under different ranking strategy

We computed a new set of mean rank precisions when sorting retrieved documents under a different ranking strategy. Both previous research (Salton 1991; Salton and Buckley 1998) and in particular our own experience with PubMed (Lu et al. 1998) suggest that ranking by relevance can result in better retrieval performance. Thus, we computed

TF-IDF scores for retrieved documents (Kim and Wilbur 2005; Lu et al. 2008) and then ranked them based on these scores. A document with a higher TF-IDF score is returned earlier in a list.

The same conclusions can be drawn with respect to the effect of using the query expansion technique based on the similar results shown in Tables 5 and 6. Superior results in mean rank precisions are seen in Table 6 compared with their counterparts in Table 5 due to a better ranking strategy. Together with experiments presented in Sect. 3.2, we confirmed an earlier expectation in (Hersh et al. 2000) where the authors believed that the difference in retrieval models (i.e. Boolean versus statistical-based) should not yield different results when evaluating query expansion techniques.

Table 6. Results of the 2006 and 2007 TREC topics with and without query expansion as measured by mean rank precision

TREC	Query expansion	P@5	P@10	P@20
2006	Yes	0.622	0.616	0.621
	No	0.613	0.592	0.584
2007	Yes	0.538	0.525	0.503
	No	0.566	0.548	0.519

Retrieved documents are sorted under the TF-IDF weighting scheme

4 DISCUSSION

4.1 PubMed's automatic term mapping process

For all the queries that we used in our work, no search tags were initially assigned. After automatic translation by PubMed's ATM, the majority of the queries were then associated with either [MeSH Terms], [Text Words] or [All Fields] as shown in Table 7. Note that there could be multiple search tags associated with a single query after translation. Take the previous *parvovirus* example, its translation included both [MeSH Terms] and [Text Word] (i.e. "parvovirus" [MeSH Terms] OR parvovirus[Text Word]).

Table 7. Results of PubMed's automatic term mapping

Table 7 Results of PubMed's automatic term mapping

Collection	Queries	[MeSH]	[Text]	[ALL]	[MeSH] OR [Text] OR [ALL]
TREC 2006	21	17	17	15	21
TREC 2007	34	28	28	23	34
Oct 17, 2005	2.4 M	1.5 M	1.4 M	1.4 M	2.0 M

The last column shows the number of queries assigned with at least one of the three search tags listed in the columns 3–5. Note that a single query can be assigned with multiple search tags (e.g. parvovirus [MeSH Terms] OR parvovirus [Text Words]). Thus for each row, the sum of the numbers in columns 3 to 5 is greater than the number in the second column (i.e. Queries). Abbreviations: [MeSH], [MeSH Terms]; [Text], [Text Words]; [ALL], [All Fields]

In order to show that the automatically generated queries are representative, we processed one day's worth of PubMed query data.4 We discarded users that issued over 50 queries/24 hours as they are likely to represent programmatic searches. For the remaining 2,657,315 queries, we obtained their corresponding PubMed translations through the Entrez Programming Utilities (eutilities) (Geer and Sayers2003). Results in the last row of Table 7 show that over 90% (2.4 M/2.65 M) of user queries are initially search tag free, over 80% (2.0 M/2.4 M) of which are expanded with at least one of the three tags studied in this work, suggesting that query expansion involving MeSH is a dominant phenomenon under real world circumstances. Other major search fields shown after translation include the ones in the following list. However, none of these search tags are related to MeSH terms. Therefore, they are beyond the scope of discussion in this paper.

1) TIAB NOT Medline [SB] (776,942): searching for terms in the title and abstract of articles in the databases that are not indexed for MEDLINE.

2) Author (380,628): searching for author names in the article.

3) Substance Name (89,510): searching for the name of a chemical in the article.

4.2 Different MeSH matching mechanisms

As we pointed out earlier in Sect. 2.3, the use of MeSH terms can be classified into two distinctive strategies:

1) If a query term is tagged with [MeSH Terms], the system will retrieve all documents indexed with this particular MeSH term as well as indexed with more specific MeSH terms beneath it in the MeSH hierarchy.

2) If a query term is tagged with either [Text Words] or [All Fields], our system will retrieve all documents indexed with one or more MeSH terms that include the query term. For instance, if a query term is *nervous system* and tagged with [Text Words] as 'nervous system' [Text Words], any documents indexedwith the MeSH Term Alcohol Abuse, Nervous SystemAlcohol Abuse, Nervous System will be retrieved in that 'nervous system' is part of this MeSH term.

To characterize the individual contribution of each of the two classes to the overall improved performance, we performed two additional experiments where only one of the two matching mechanisms was exploited. Results are compared to each other in Table 8, and are also compared to the cases when none or both matching mechanisms were used. Results in Table 8 show that directly matching the MeSH term and its more specific terms is the major contribution for performance improvement in TREC 2007 while matching words in MeSH terms contributed significantly in TREC 2006. Therefore, we are unable to draw a conclusion on which one is more important for improving performance based on these experiments.

Table 8. Performance (in F-measure) comparisons using different MeSH matching strategies

TREC	None	[MeSH Terms]	[Text Words] or [All Fields]	Both
2006	0.334	0.346	0.407	0.406
2007	0.214	0.260	0.236	0.264

Experiments were performed on both TREC 2006 and 2007 data

As a separate effort, we performed an additional experiment where we turned off the automatic MeSH explosion feature (i.e. more specific MeSH terms are no longer searchable) in the context of evaluating this specific feature. When compared to the results in the third column of Table 8, we see a performance drop from 0.346 to 0.334 in TREC 2006, and from 0.260 to 0.212 in TREC 2007, respectively. This leads us to conclude that the MeSH automatic explosion feature plays a critical role in the first type of MeSH usage.

4.3 Comparison with PubMed output

The main objective of this work is to discover useful features to improve PubMed search performance. In spite of the fact that our analysis was not based on output directly from PubMed, the retrieved document set by our custom search system was highly consistent with the actual PubMed output as shown in Figure 3. The F-measures are almost identical for the systems. The slight discrepancies in performance come from two sources in our analysis: (a) we limited our searches with only three search field tags; and (b) for the search tag [all fields], our search was limited to text words in title, abstract, and MeSH terms. Any other fields (e.g. author names) were not considered in our computation. The minor performance increase when using the custom-built system suggests (a) the three search tags that we selected to evaluate are critical for PubMed; and (b) that perhaps other search field tags could not further help to retrieve relevant documents for our query set.

5 CONCLUSIONS AND FUTURE WORK

Based on the results of our large-scale analysis comprised of the 55 real biological questions and independently judged relevant documents, we conclude that query expansion using MeSH through PubMed's Automatic Term Mapping process can generally result in retrieving more relevant documents. However, this type of improvement may not prove useful for those users looking only at the top ranked returned documents (e.g. the first 20 returned documents).

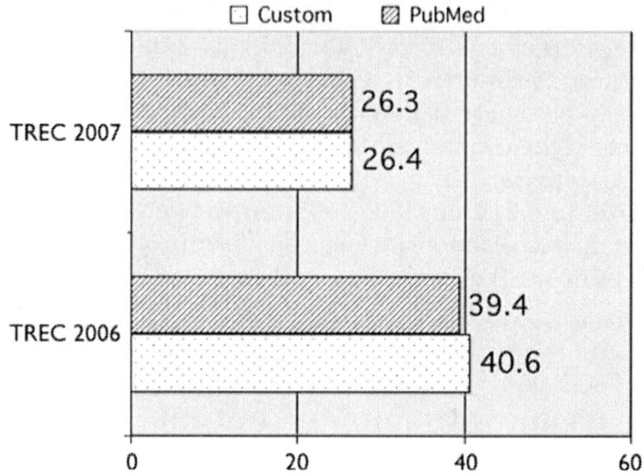

Figure 3. The comparisons of F-measures based on outputs from PubMed vs. from our custom search system

The comparison results are useful in suggesting changes in PubMed (e.g. the display order). However, as we pointed out earlier, this work makes two assumptions. One assumption is that the automatically generated queries represent real user queries. The other assumption is that the documents in the relevant set are the ground truth and there are no more relevant documents in the TREC set. Our future research goal is to address both issues by involving human experts in evaluations.

FOOTNOTES

[1]http://highwire.stanford.edu/.

[2]Any changes to PubMed's ATM after March 2008 by the National Library of Medicine may result in slightly different translations as shown.

[3]http://www.ncbi.nlm.nih.gov/books/bv.fcgi?rid=helppubmed.chapter.pubmedhelp.

[4]ftp://ftp.ncbi.nlm.nih.gov/toolbox/pubmed/query-logs/README.

ACKNOWLEDGMENTS

This research was funded by the Intramural Research Program of NIH, National Library of Medicine. The authors are grateful to the TREC organizers for their efforts in producing and making the text collection and relevance judgments publicly available.

REFERENCES

1. Abdou, S., Ruck, P., & Savoy, J. (2005). Evaluation of stemming, query expansion and manual indexing approaches for the genomics track. In *Proceedings of the Fourteenth Text REtrieval Conference, Gaithersburg, MA*. Gaithersburg, MD: Department of Commerce, National Institute of Standards and Technology.

2. Aronson, A. R. (1996). The effect of textual variation on concept based information retrieval. In *Proceedings of the AMIA Annual Fall Symposium* (pp. 373–377). Bethesda, MD: American Medical Informatics Association.

3. Aronson, A. R., & Rindflesch, T. C. (1997). Query expansion using the umls metathesaurus. In *Proceedings of the AMIA Annual Fall Symposium* (pp. 485–489). Bethesda, MD: American Medical Informatics Association.

4. Bacchin, M., & Melucci, M. (2005). Symbol-based query expansion experiments at TREC 2005 genomics track. In*Proceedings of the Fourteenth Text REtrieval Conference*. Gaithersburg, MD: Department of Commerce, National Institute of Standards and Technology.

5. Carlin, B. G. (2004). Pubmed automatic term mapping. *Journal of the Medical Library Association, 92*(2), 168.

6. Gault, L. V., Shultz, M., & Davies, K. J. (2002). Variations in medical subject headings (mesh) mapping: From the natural language of

patron terms to the controlled vocabulary of mapped lists. *Journal of the Medical Library Association, 90*(2), 173–180.

7. Geer, R. C., & Sayers, E. W. (2003). Entrez: Making use of its power. *Briefings in Bioinformatics, 4*(2), 179–184.

8. Guo, Y., Harkema, H., & Gaizauskas, R. (2004). Sheffield university and the TREC 2004 gemonics track: Query expansion using synonymous terms. In *Proceedings of the Thirteenth Text REtrieval Conference*. Gaithersburg, MD: Department of Commerce, National Institute of Standards and Technology.

9. Hersh, W. (2003). *Information retrieval: A health and biomedical perspective* (2nd ed.). New York, NY: Springer-Verlag.

10. Hersh, W. & Bhupatiraju, R. T. (2003). TREC Genomics track overview. *The Twelfth Text Retrieval Conference, TREC 2003* (pp. 14–23). Gaithersburg, MD: Department of Commerce, National Institute of Standards and Technology.

11. Hersh, W., Bhupatiraju, R. T., & Price, S. (2003). Phrases, boosting, and query expansion using external knowledge resources for genomic information retrieval. In *Proceedings of The Twelfth Text REtrieval Conference.* Gaithersburg, MD: Department of Commerce, National Institute of Standards and Technology.

12. Hersh, W., Bhupatiraju, R. T., Ross, L., Johnson, P., Cohen, A., & Kraemer, D. (2006). TREC 2006 genomics track overview. In *Proceedings of the Thirteenth Text REtrieval Conference*. Gaithersburg, MD: Department of Commerce, National Institute of Standards and Technology.

13. Hersh, W., Cohen, A. M., Roberts, P., & Rekapalli, H. K. (2006). TREC 2006 genomics track overview. In*Proceedings of the Fifteenth Text REtrieval Conference.* Gaithersburg, MD: Department of Commerce, National Institute of Standards and Technology.

14. Hersh, W., Cohen, A. M., Roberts, P., & Rekapalli, H. K. (2007). TREC 2007 genomics track overview. In*Proceedings of the Sixteenth Text

REtrieval Conference. Gaithersburg, MD: Department of Commerce, National Institute of Standards and Technology.

15. Hersh, W., Price, S., & Donohoe, L. (2000). Assessing thesaurus-based query expansion using the umls metathesaurus. In *Proceedings of AMIA Annual Symposium* (pp. 344–348). Bethesda, MD: American Medical Informatics Association.
16. Herskovic, J. R., Tanaka, L. Y., Hersh, W., & Bernstam, E. V. (2007). A day in the life of PubMed: Analysis of a typical day's query log. *Journal of the Medical Library Association, 14*(2), 212–220.
17. Kim, W., & Wilbur, W. J. (2005). A strategy for assigning new concepts in the MEDLINE database. In*Proceedings of AMIA Annual Symposium* (pp. 395–399). Bethesda, MD: American Medical Informatics Association.
18. Lindberg, D., Humphreys, B., & Mccray, A. (1993). The unified medical language system. *Methods of Information in Medicine, 32*(4), 281–291.
19. Lu, Z., Kim, W., & Wilbur, W. J. (2008). Evaluating relevance ranking strategies for medline retrieval. *Journal of the American Medical Informatics Association*, in press.
20. Mitra, M., Singhal, A., & Buckley, C. (1998). Improving automatic query expansion. In *SIGIR '98: Proceedings of the 21st Annual International ACM SIGIR Conference on Research and Development in Information Retrieval* (pp. 206–214). New York, NY, USA: ACM.
21. Noreen, E. (1989). *Computer Intensive Methods for Testing Hypotheses: An Introduction*. Hoboken, NJ: John Wiley & Sons, Inc.
22. Robertson, S. E., & Walker, S. (1994). Some simple effective approximations to the 2-poisson model for probabilistic weighted retrieval. In *SIGIR '94: Proceedings of the 17th Annual International ACM SIGIR Conference on Research and Development in Information Retrieval* (pp. 232–241). New York, NY, USA. New York: Springer-Verlag.

23. Salton, G. (1991). Developments in automatic text retrieval. *Science, 253*(5023), 974.

24. Salton, G., & Buckley, C. (1988). Term weighting approaches in automatic text retrieval. *Information Processing & Management, 24,* 513–523.

25. Salton, G., & Buckley, C. (1997). Improving retrieval performance by relevance feedback. In *Readings in Information Retrieval* (pp. 355–364). San Francisco, CA, USA: Morgan Kaufmann.

26. Shultz, M. (2006). Mapping of medical acronyms and initialisms to medical subject headings (mesh) across selected systems. *Journal of the Medical Library Association, 94*(4), 410–414.

27. Smith, A. M. (2004). An examination of pubmed's ability to disambiguate subject queries and journal title queries.*Journal of the Medical Library Association, 92*(1), 97–100.

28. Srinivasan, P. (1996). Query expansion and MEDLINE. *Information Processing & Management, 32*(4), 431–443.

29. Wilbur, W. (1994). Non-parametric significance tests of retrieval performance comparisons. *Journal of Information Science, 20*(4), 270–284.

30. Wilbur, W. J., & Sirotkin, K. (1991). The automatic identification of stop words.*Journal of Information Science, 18*(1992), 45–55.

3

ON INFORMATION RETRIEVAL METRICS DESIGNED FOR EVALUATION WITH INCOMPLETE RELEVANCE ASSESSMENTS

Tetsuya Sakai[1], Noriko Kando[2]

[1]NewsWatch, Inc., Tokyo, Japan

[2]National Institute of Informatics, Tokyo, Japan

ABSTRACT

Modern information retrieval (IR) test collections have grown in size, but the available manpower for relevance assessments has more or less remained constant. Hence, how to reliably evaluate and compare IR systems using*incomplete* relevance data, where many documents exist that were never examined by the relevance assessors, is receiving a lot of attention. This article compares the robustness of IR metrics to incomplete relevance assessments, using four different sets of graded-relevance test collections with submitted runs—the TREC 2003 and 2004 robust track data and the NTCIR-6 Japanese and Chinese IR data from the crosslingual task. Following previous work, we artificially reduce the original relevance data to simulate IR evaluation environments with extremely incomplete relevance data. We then investigate the effect of this reduction on *discriminative power*, which we define as the proportion of system pairs with a statistically significant difference for a given probability of Type I Error, and on *Kendall's rank correlation*, which reflects the overall resemblance of two system rankings according to two different metrics or two different relevance data sets. According to these experiments, Q′, nDCG′ and AP′ proposed by Sakai are superior to bpref proposed by Buckley and

Voorhees and to Rank-Biased Precision proposed by Moffat and Zobel. We also point out some weaknesses of bpref and Rank-Biased Precision by examining their formal definitions.

KEYWORDS

Evaluation metrics; Relevance assessments; Test collections; Incompleteness

1 INTRODUCTION

An information retrieval (IR) test collection comprises a document collection, a set of search requests and a set of manually judged relevant documents for each request. Following TREC 1 parlance, hereafter the search requests will be referred to as *topics*, and the set of relevant documents will be referred to as *qrels*. The methodology of using a test collection for evaluating and comparing IR techniques was established in the 1960s, through the Cranfield 2 Test by Cleverdon (1967). Since then, laboratory experiments using test collection have always played a central role for the progress of IR techniques, as they are objective, efficient and repeatable (Voorhees 2002).

Now, in the twenty-first century, evaluation using test collections is still a necessity for most IR researchers. However, the 1990s actually saw a major departure from the original Cranfield 2 experiments, with the advent of TREC, NTCIR 2 and CLEF3 evaluation efforts for constructing very large-scale test collections. On the surface, the main difference between the Cranfield 2 test collection and the modern test collections is the document collection size: The Cranfield 2 test collection contained only 1,400 documents; The TREC and NTCIR test collections typically contain between half a million to one million documents. However, a more important difference is that while the small scale Cranfield 2 collection had*complete* relevance assessments, the modern test collections do not: It is simply not feasible to examine the documents exhaustively for these large scale collections.

For creating the qrels, TREC, CLEF and NTCIR all adopt a mechanism called*pooling*, which works as follows:

1) Participants submit their "runs" (collections of ranked lists for each topic, where each ranked list usually contains up to 1,000 documents) to the organisers;

2) For each topic, organisers take the top *k* (typically 100) documents from some of the submitted runs and obtain a list of unique documents, i.e, a document pool;

3) For each topic, assessors judge the relevance of all documents within the pool.

Hence, despite the fact that pooling is a very efficient way of collecting relevant documents, qrels formed through pooling are possibly *incomplete*. That is, there may exist relevant documents within the document collection, which none of the participating systems managed to retrieve and therefore are outside the qrels (Voorhees 2002).

While the collection sizes tend to grow monotonically in order to mimic real-world data such as the Web, the available manpower for relevance assessments remains more or less constant, and therefore test collections are destined to become more and more incomplete. Using an incomplete test collection for IR evaluation raises the following concerns at least:

a) Is it possible to reliably compare two participating systems and judge which is superior? Were the test collection less incomplete, would this judgement be the same?

b) Is it possible to reliably compare a participating system and a new system that never contributed to the pool? How about two new systems? In short, is the incomplete test collection*reusable*?

For these reasons, IR evaluation using incomplete relevance assessments is receiving more attention than ever.

One obvious approach to tackling these problems is to devise IR effectiveness metrics that are *robust* to relevance data incompleteness: We say that an IR metric is robust to incompleteness if system comparison results based on an incomplete set of relevance data are similar to those based on a less incomplete one. This article follows this approach, and more specifically, addresses the issues mentioned in (a) above. Below, we discuss three existing studies that are directly related to the present one. We shall discuss other related work in Sect. 2.

Buckley and Voorhees (2004) proposed an IR evaluation metric called *bpref* (binary preference) which is highly correlated with Average Precision (AP) when full relevance assessments are available and is yet more robust when the relevance assessments are reduced. Recent TREC tracks have used this metric along with AP. Bpref penalises a system if it ranks a *judged nonrelevant* document above a judged relevant one, and is indepedendent of how the *unjudged* documents are retrieved.

More recently, Moffat et al. (2007) introduced an IR evaluation metric called Rank-Biased Precision (RBP) which they claimed is suitable for evaluation with incomplete relevance data. RBP assumes that the probability that the user moves from a document at Rank r to Rank $(r + 1)$ is a constant p, regardless of the relevance (level) of the document at Rank r. As it does not have a recall component, adding more relevant documents to the qrels always increases the RBP score.

Sakai (2007a) reported that applying Q-measure (or simply, Q), AP and normalised Discounted Cumulative Gain (nDCG) to a *condensed list*, i.e., a ranked list of documents obtained by removing all unjudged documents from the original list, is a simpler and a better solution than bpref for handling relevance data incompleteness. The metrics applied to condensed lists will hereafter be referred to as Q′, AP′ and nDCG′, respectively.

This article compares the robustness of Q(′), AP(′), nDCG(′), bpref and RBP to incomplete relevance assessments, using four different sets of *graded-relevance*test collections with submitted runs—the TREC 2003 and 2004 robust track data and the NTCIR-6 Japanese and Chinese IR data from the crosslingual task. We believe that evaluating IR systems using graded relevance is important for the progress of IR because, if one adheres to IR evaluation based on binary relevance, it would be very difficult for him to devise an IR algorithm that can retrieve highly relevant documents on top of partially relevant ones. Following previous work, we artificially reduce the original relevance data to simulate IR evaluation environments with extremely incomplete relevance data. We then investigate the effect of this reduction on *discriminative power* (Sakai 2006b, 2007b), which we define as the proportion of system pairs with a statistically significant difference for a given probability of Type I Error, and on *Kendall's rank correlation* (Voorhees2001)), which reflects the overall resemblance of two system rankings according to two different metrics or two different qrels. According to these experiments, Q′, nDCG′ and AP′ are superior to bpref and RBP.

This article generalises a recent study by Sakai (2007a), in that (1) While he used the NTCIR-3 and NTCIR-5 Japanese/Chinese data, we use TREC robust track data and NTCIR-6 Japanese/Chinese data to obtain more general and substantial conclusions; (2) We compare RBP with the other metrics, after discussing some properties of the metrics that immediately follow from their formal definitions.

The remainder of this article is organised as follows. Section 2 provides an overview of related studies. Section 3 defines and discusses the characteristics of AP(′), Q(′), nDCG(′), bpref and RBP. Section 4 describes the TREC and NTCIR data we used for comparing the robustness of these metrics to relevance data incompleteness. Section 5 compares the metrics in terms of discriminative power based on statistical significance tests. Section 6 compares the metrics in terms of Kendall's rank correlation between the entire system rankings. Finally, Sect. 7concludes this article.

2 RELATED WORK

This section provides an overview of previous work related to the present study.

There are at least two approaches to tackling the relevance data incompleteness problem: One is to try to construct a better test collection more efficiently, and another is to devise or choose reliable IR metrics, *given* a test collection. Along the first line of research, methods for creating judgment pools efficiently were proposed by Cormack et al. (1998) and by Zobel (1998). Soboroff et al. (2001) proposed a method for ranking systems without any relevance assessments, but subsequently Aslam and Savell (2003) pointed out that the method tends to rank them by "popularity" rather than performance. More recently, Carterette et al. (2006) analyzed the distribution of AP over all possible assignments of relevance to all unjudged documents and proposed a method to construct a test collection with minimal relevance assessments; Büttcher et al. (2007) proposed to "expand" existing relevance assessments by treating them as training data for machine learning.

This article takes the latter approach, of choosing reliable IR metrics for handling relevance data incompleteness. The proposals of the aforementioned bpref (Buckley and Voorhees 2004), RBP (Moffat et al.

2007) and Q′, nDCG′ and AP′ (Sakai 2007a) fall into this category. Also along this line, Aslam et al. (2006) and Yilmaz and Aslam (2006) proposed *Induced AP*, *Subcollection AP* and *Inferred AP*. Induced AP is exactly what we call AP′. We do not consider Subcollection AP and Inferred AP in our present study, because (1) While the goal of Yilmaz and Aslam was to estimate the true AP values, ours is not: We prefer to explore different metrics, especially those that can handle graded relevance; (2) Both Subcollection AP and Inferred AP require knowledge of the *pooled but unjudged* documents, which limits their applicability; 4 (3) According to Bompada et al. (2007), Inferred AP is not as robust as the original nDCG for evaluation with incomplete relevance data.

Another metric proposed for handling incomplete relevance data, called *RankEff* (Grönqvist 2005), has been examined by Büttcher et al. (2007). However, Sakai (2008b) points out that RankEff is in fact equivalent to an existing variant of bpref called *bpref_N*, also known as "bpref_allnonrel" implemented in trec_eval, the standard IR evaluation software for TREC. Sakai (2007a) showed both analytically and empirically that bpref_N is not a good evaluation metric. See Sect. 3.2 for more discussions.

Büttcher et al. (2007) also used Precision at *l judged* documents, which relies on condensed lists just like Q′, AP′ and nDCG′. However, Precision is not a satisfactory metric for us because: (1) It ignores the ranks of retrieved relevant documents; (2) It does not average well, especially with a large document cut-off; (3) With a small document cut-off, it gives unreliable results as systems are evaluated based on a small number of observations, i.e., documents near the top of the ranked list (Sakai 2007f).

De Beer and Moens (2006) proposed *rpref*, a graded-relevance version of bpref. Sakai (2007a) pointed out that it has minor bugs, and proposed *rpref_relative2* by fixing them. However, he reported that it does not have any advantage over Q′, AP′ and nDCG′ despite its complexity.

Sakai (2007c) conducted a study similar to the present one, but focussed on the task of finding one highly relevant document. He showed that the application of Reciprocal Rank, O-measure (Sakai 2006c) and P+-measure (Sakai 2006a, 2007e) to condensed lists is an effective way of handling the relevance data incompleteness problem.

3 FORMAL DEFINITIONS OF THE IR METRICS

This section formally defines the IR metrics we consider, namely, Q('), AP('), nDCG('), bpref and RBP, and also discusses their properties that immediately follow from the definitions. Among these metrics, only Q('), nDCG(') and RBP can handle graded relevance.

3.1 Q, AP and nDCG

Let LL denote a *relevance level*, and let gain(L)gain(L) denote the *gain value* for retrieving an LL-relevant document for a particular topic. Following the NTCIR tradition (Kando 2007), this article assumes that we have S-relevant (i.e., highly relevant), A-relevant (i.e., relevant) and B-relevant (i.e., partially relevant) documents. Other documents, i.e., judged nonrelevant documents and unjudged documents, are considered nonrelevant and therefore do not carry a gain value. We let $gain(S) = 3$, $gain(A) = 2$ and $gain(B) = 1$ hereafter as metrics such as Q and nDCG are robust to the choice of gain values (Sakai 2007f). As for the TREC data, which only have "highly relevant" and "relevant" documents, we treat the former as S-relevant, and the latter as B-relevant. The latter were treated as B-relevant rather than A-relevant because it has been reported that there are many marginally or partially relevant documents in the TREC qrels: Sormunen (2002) reported that about one half of their TREC qrels were only marginally relevant; Sakai and Sparck Jones (2001) reported that only about 56% of the TREC qrels were highly relevant for a subcollection of the early TREC document sets.

Let $R(\mathcal{L})$ denote the number of $\mathcal{L}$-relevant documents, and let $R = \sum_{\mathcal{L}} R(\mathcal{L})$. Let $cg(r) = \sum_{1 \leq i \leq r} g(i)$ denote the *cumulative gain* at Rank r of the system output, where $g(i) = gain(\mathcal{L})$ if the document at Rank i is $\mathcal{L}$-relevant and $g(i) = 0$ otherwise (i.e., if the document at Rank i is either judged nonrelevant or unjudged). Let $cg_I(r)$ denote the cumulative gain of an *ideal* ranked output, where an ideal ranked output is one that satisfies $g(r) > 0$ for $1 \leq r \leq R$ and $g(r) \leq g(r-1)$ for $r > 1$. For NTCIR, for example, listing up all S-relevant documents, followed by all A-relevant documents, followed by all B-relevant documents produces an ideal ranked output. Note that whether "nonrelevant" (i.e., either judged nonrelevant or unjudged) documents are retrieved below these relevant documents does not matter, as the nonrelevant documents do not carry gain values. Moreover, note that several "ideal" ranked outputs can exist in general, since documents can be interchanged within each relevance level.

Let *isrel*(*r*) be one if the document at Rank *r* is relevant and zero otherwise, and let $count(r) = \sum_{1 \leq i \leq r} isrel(i)$. Clearly, precision at Rank *r* is given by *P*(*r*)=*count*(*r*)/*r*.

We first define Q-measure:

$$Q\text{-measure} = \frac{1}{R}\sum_{r} isrel(r)BR(r) \quad (1)$$

$$BR(r) = \frac{\beta cg(r) + count(r)}{\beta cg_I(r) + r} \quad (2)$$

where *BR*(*r*) is called the *blended ratio* and β is a *persistence* parameter. Because*BR*(*r*) has an *r* in the denominator (just like *P*(*r*)), Q-measure is guaranteed to become smaller as a relevant document goes down the ranked list. A large β (e.g., β = 100) alleviates this effect, and makes Q-measure more forgiving for relevant documents near the bottom of the ranked list. Conversely, a small β (e.g., β = 1) imposes more penalty. Sakai (2007d) showed empirically that β = 1,10 are good choices, so we take β = 1 throughout this article. Note also that β = 0 reduces Q-measure to AP:

$$AP = \frac{1}{R}\sum_{r} isrel(r)\frac{count(r)}{r} = \sum_{r} isrel(r)P(r). \quad (3)$$

For a given logarithm base *a*, let the *discounted* gain at Rank *r* be *dg*(*r*) = *g*(*r*)/log$_a$(*r*) for *r* > *a* and dg(r)=g(r) for *r* ≤ *a*. Similarly, let $dg_I(r)$ denote the discounted gain for an ideal ranked list. nDCG at document cut-off *l* is defined as:

$$nDCG_l = \sum_{1 \leq r \leq l} dg(r) / \sum_{1 \leq r \leq l} dg_I(r). \quad (4)$$

Throughout this article, we let *l* = 1,000 as it has been reported that nDCG with a small document cut-off is unreliable (Sakai 2007f). Moreover, we let *a* = 2 because it has been reported that nDCG with a large logarithm base is counterintuitive and lacks discriminative power (Sakai 2007d), despite the fact that this parameter was designed to reflect persistence just like RBP's *p* and Q-measure's β. We shall come back to this issue in Sect. 3.4.

3.2 Q′, AP′, nDCG′ and bpref

Sakai (2007a) reported that Q′, AP′ and nDCG′ are simpler and better solutions to the problem of evaluating IR systems with incomplete relevance data than *bpref* (Buckley and Voorhees 2004). Recall that these represent the application of Q, AP and nDCG to condensed lists, respectively.

Let r' denote the rank of a document in a condensed list, whose original rank was$r(\geq r')$. Let N denote the number of *judged nonrelevant* documents. Then *bpref* can be expressed as follows (Sakai 2007a):

$$bpref = \frac{1}{R}\sum_{r'} isrel(r')\left(1 - \frac{\min(R, r' - count(r'))}{\min(R, N)}\right) \tag{5}$$

where r' - $count(r')$ is the number of judged nonrelevant documents ranked above the relevant one at Rank r', or the *misplacement penalty* with respect to this particular relevant document. Clearly, for any topic such that $R \leq N$, bpref reduces to:

$$bpref\_R = \frac{1}{R}\sum_{r'} isrel(r')\left(1 - \frac{\min(R, r' - count(r'))}{R}\right). \tag{6}$$

In fact, $R \leq N$ in holds for all topics used in our experiments (See also Table 2), so*bpref is always bpref_R* in our study.

Sakai (2007a) pointed out that the only essential difference between AP′ and bpref is that, while the former uses r' for scaling each misplacement penalty r' - $count(r')$, the latter uses a constant (e.g., R). Compare Eq. 6 with

$$AP' = \frac{1}{R}\sum_{r'} isrel(r')\frac{count(r')}{r'} \tag{7}$$

$$= \frac{1}{R}\sum_{r'} isrel(r')\left(1 - \frac{r' - count(r')}{r'}\right). \tag{8}$$

Scaling by a constant is generally not good, especially if the constant is large, because this means that the misplacement penalties with respect to the top ranked relevant documents are virtually ignored (Sakai 2007a). For example, suppose that there is a condensed list that has a judged nonrelevant document at Rank 1 and a relevant document at Rank 2. For this document at rank $r' = 2$, the misplacement penalty is $r'-count(r')=2-1=1$, and $P(r')=1/2$. Thus, the existence of the judged nonrelevant document at Rank 1 weighs heavily in the case of AP′AP′. In contrast, this has very little impact on bpref, because the misplacement penalty is divided not by $r' = 2$ but by a large number, namely, R or N. In other words, bpref lacks the "top heaviness" of AP′, which is one of the main strengths of the original AP. It is clear that bpref_N (Sakai 2007a), which always uses N for scaling the misplacement penalty, suffers severely from this problem, as N is generally a very large number: See, for example, Table 2 which we shall discuss later. Sakai (2007a) showed experimentally that bpref_N indeed performs very poorly.

It should be noted that AP′ is actually implemented in trec_eval. However, it appears that it was never properly examined until Yilmaz and Aslam (2006) and Sakai (2007a) rediscovered it.

3.3 RBP

We now formally define RBP (Moffat et al. 2007; Moffat and Zobel 2008). Let HHdenote the highest relevance level across all topics. RBP can be expressed as follows:

$$RBP = \frac{1-p}{gain(\mathcal{H})} \sum_{r} g(r) p^{r-1} \tag{9}$$

where $p(\leq 1)$ is a persistence parameter. A high value of p represents a persistent user; a low value represents an impatient one. As Moffat and Zobel (2008) explored p = 0.5, 0.8, 0.95, we start our own experiments with the same values, denoting each version of RBP by RBP.5, RBP.8 and RBP.95. In all of our experiments, we let gain (H)=gain(S)=3. Recall that our NTCIR data have S-, A- and B-relevant documents, but our TREC data have S- and B-relevant documents only.

The assumption behind RBP is that the user, after examining the document at Rank r, will examine the document at Rank $(r + 1)$ with probability p or stop scanning the ranked list with probability $1 - p$. Thus the model assumes that the transition probability is independent of the relevance of the document at Rank r. Whether this assumption is realistic or not is debatable, but this does make RBP easy to interpret and to compute. Moreover, Moffat et al. (2007) argue that RBP is suitable for evaluation with incomplete relevance data as it is guaranteed to increase as more relevance judgments are added (since it does not have a recall component) and the error due to unjudged documents can be quantified.

However, we can discuss RBP's possible weaknesses. Firstly, RBP may give a very low score even to an ideal ranked output: In fact, the fact that it does not rely on recall implies that it denies the very existence of an "ideal" ranked output. From Eq. 9, it is clear that the RBP for an ideal ranked list in a binary relevance environment equals $(1-p)\sum_{r=1}^{R} p^{r-1}$. Table 1 shows the RBP value for an ideal ranked output for p = 0.5, 0.8, 0.95 and R = 1, 10, 100, 1,000. When p = 0.95, for example, an ideal ranked output for a topic with R = 10 receives an RBP of .4013, while one for a topic with R = 100 receives .9941. Whether it is good to average such a measurement across topics is debatable, but it is at least a fact that topics with many relevant documents can have a far larger impact on Mean RBP than those with few relevant ones.

Table 1. Values of RBP for an ideal ranked output

	RBP.5	RBP.8	RBP.95
$R = 1$	.5	.2	.05
$R = 10$	.9990	.8926	.4013
$R = 100$	1	1	.9941
$R = 1{,}000$	1	1	1

Moreover, Table 1 shows the extreme cases of when R = 1: it can be observed that the RBP of an ideal ranked output (i.e., one that has the only one relevant document at Rank 1) can range from 0.05 (p = 0.95) and 0.5 (p = 0.5), since RBP in this case equals 1 – p. Thus the user's persistence, i.e., the probability of moving from a document from Rank r to that at Rank (r + 1), influences the effectiveness value of the same ranked output quite drastically, even though only the document at Rank 1 is being examined. Whether this is a desirable feature for an IR metric is also debatable. In contrast, Q, nDCG and AP are by definition guaranteed to be one whenever the system output is an ideal ranked output regardless of the value of R, since they are based on comparing the system output with the ideal one.5

We further argue that depending on recall is not necessarily bad. The real user may have some idea of the number of relevant documents, due to his background knowledge, or if not, by looking at the total number of hits shown in the IR interface. Moreover, even if this is not the case, a good IR performance metric is not necessarily one that closely mimics "user satisfaction." For example, a user may be very satisfied with the ranked output, having found a decent document, but he may have missed ten other documents that are in fact more relevant than the one he has found. That is, *the user may be happy, just because he is ignorant.* From a conscientious system developer's point of view, if he *knows* that there are ten relevant documents that should be retrieved, then he would design a system that can retrieve as many of them as possible rather than a system that makes the user "happy" by showing just one relevant document and hiding the other relevant ones completely. Hence Q and AP depend directly on R, the number of judged relevant documents, and even nDCG depends on it indirectly, as it relies on an ideal ranked output which lists up all relevant documents.

3.4 Top-heaviness of RBP, AP, Q and nDCG

Figure 1 compares the "top-heaviness" of RBP, AP, Q and nDCG, by considering a ranked output that contains exactly one relevant document, and making it move from Rank 1 to Rank 20. The graph at the top shows the situation when $R = 10$, and the one at the bottom shows the situation when $R = 100$, both under a binary relevance environment. Note that the three RBP curves are not affectd by the value of R. From the figure, it can be observed that RBP.5 is probably too top-heavy: it basically ignores any relevant document retrieved below Rank 10. This makes evaluation very unstable, as we shall see in our experiments in Sect. 5. RBP.8 gives a reasonable "rank bias": RBP.95 looks almost like a straight line, compared to other metrics such as Q-measure and nDCG.

In Figure 1, the top-heaviness curve of AP is almost completely hidden by that of Q-measure, because in a binary relevance environment, *Q-measure* = *AP* holds if there is no relevant document below Rank R, while *Q-measure* > *AP* holds if there is at least one relevant document below Rank R (Sakai 2006c). Thus the AP curve actually begins to deviate from the Q-measure one at Rank 11 in the graph at the top (where $R = 10$).

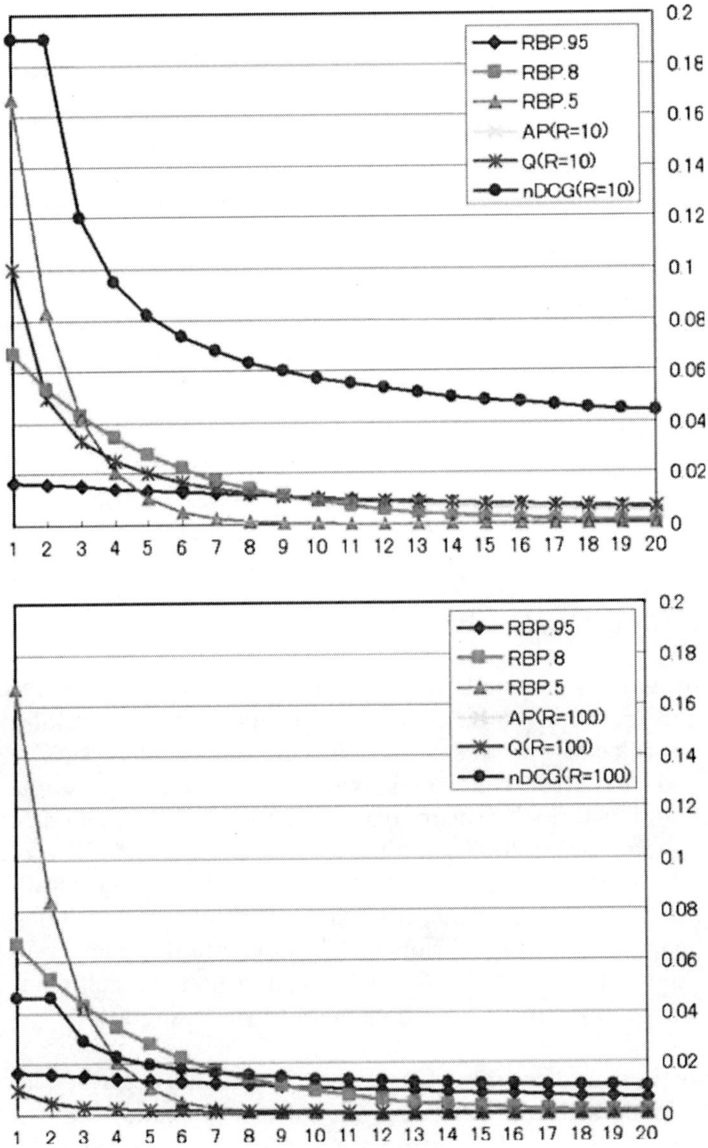

Figure 1. Comparison of "top-heaviness"

It can also be observed that the top-heaviness curves of nDCG have a minor problem: nDCG with a logarithm base of 2 cannot distinguish between a system that has a relevant document at Rank 1 and one that has a relevant document at Rank 2. This is because, according to the original definition of nDCG (which we stick to), *gain discounting* cannot be applied to ranks above a(=2). This is precisely why using a large a with nDCG is no good (Sakai 2007d): it makes the top-heaviness curve even flatter.6 It should also be noted that the top-heaviness curve for *nCG* (Järvelin and Kekäläinen 2002; Kekäläinen 2005), the undiscounted version of nDCG, is a completely flat line. That is, to nCG, it does not matter at all at which rank the relevant document is found. This explains why nCG performs very poorly: and the same goes for Precision at l (Sakai 2007f).

To sum up, the IR metrics we consider in this study all have a mechanism, each in its own way, of penalising relevant documents found near the bottom of the ranked list. But the graphs suggest that using p = 0.5 for RBP may not be good for reliable evaluation. This we will verify in our experiments described below.

4 FULL AND REDUCED DATA

Table 2 provides some statistics of the TREC and NTCIR data we used for evaluating the IR metrics for the purpose of evaluation with incomplete relevance assessments. We chose these data sets as we wanted "ad hoc" test collections with graded relevance data. The "TREC03" and "TREC04" data are from the TREC 2003 and 2004 robust track (Voorhees 2004; 2005), and the "NTCIR-6J" and "NTCIR-6C" data are from the NTCIR-6 crosslingual track (Kando 2007). The TREC runs are English monolingual runs, and the NTCIR-6J runs include both monolingual and crosslingual runs for the Japanese document retrieval subtask. Similarly, the NTCIR-6C runs include both monolingual and crosslingual runs for the Chinese document retrieval subtask.

For conducting our discriminative power experiments described in Sect. 5, we randomly selected one run from each participating team. Thus, with the TREC03 data, for example, we used 16 runs, which yields 16*15/2 = 120 combinations of teams for significance testing. Figure 2 shows the distribution of AP values over the runs thus selected. For computing Kendall's rank correlation, we wanted more runs, and we also wanted the same number of runs across all four data sets.7 We therefore randomly sampled 30 runs from each data set, disregarding which team each run comes from.

Table 2. TREC and NTCIR data used in our experiments

	TREC03	TREC04	NTCIR-6J	NTCIR-6C
#Topics	50	49	50	50
#Documents	Approx. 529,000		858,400	901,446
Pool depth	125	100	100	100
Average N	925.5	654.6	1157.9	999.4
Range N	[292, 2050]	[132, 1371]	[480, 2732]	[414, 1907]
Average R	33.2	41.2	95.3	88.1
Range R	[4, 115]	[3, 161]	[4, 311]	[15, 400]
S-relevant	8.1	12.5	2.5	21.6
A-relevant	–	–	61.1	30.4
B-relevant	25.0	28.8	31.7	36.1
#Teams	16	14	12	11
#All runs	78	110	73	45
#Runs used for rank correlation	30	30	30	30

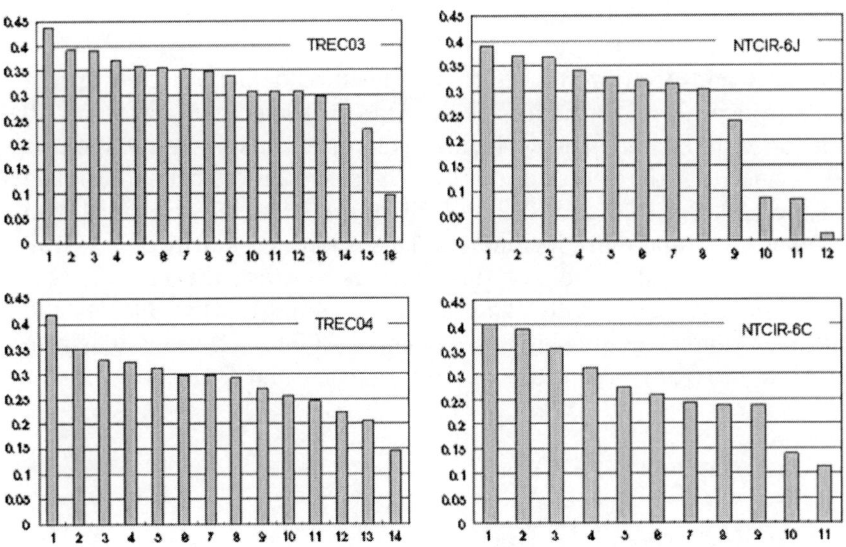

Figure 2. AP values of the participating teams

To examine the effect of relevance data incompleteness on the IR metrics, we created *reduced relevance data* from the full relevance data, following the original methodology by Buckley and Voorhees (2004): First, for each topic, we created a randomised list of judged relevant documents of size *R*, and a separate randomised list of judged nonrelevant documents of size *N*. Then, for each*reduction rate* $j \in \{90, 70, 50, 30, 10\}$, we created a reduced set of relevance data by taking the first R_j and N_j documents from the two lists, respectively,where Rj=max(1,truncate(R∗j/100)) and Nj=max(10,truncate(N∗j/100)). The contents 1 and 10 have been copied from Buckley and Voorhees (2004), representing the minimum number of judged relevant and nonrelevant documents required for a topic, respectively. In practice, the constant 10 was seldom used since *N* was generally very large. This stratified sampling is essentially equivalent to random sampling from the entire set of judged documents (Yilmaz and Aslam2006).

The above method of random sampling from the original qrels may be criticised: Possibly, a better method of studying the effect of incompleteness would be to use the actual pools for each topic, and vary the pool depth. However, we prefer to be faithful to the methodology by Buckley and Voorhees as one of the main goals of this study is to contrast their claims regarding bpref (Buckley and Voorhees 2004) with our new findings. We will report on our "shallow pool" experiments elsewhere (Sakai 2008a).

Figures 3 and 4 show the effect of relevance data reduction on the absolute overall performances (e.g., Mean AP) averaged across all 30 runs for each data set. The horizontal axis represents the reduction rate *j*. It is clear that the values of the metrics based on the original ranked lists, i.e., AP, Q, nDCG and RBP, quickly diminish as the relevance data becomes more and more incomplete. This is not necessarily a flaw: RBP has been designed to behave this way. In contrast, the bpref_R (i.e., bpref) curve is relatively flat, and this much supports what Buckley and Voorhees reported (Buckley and Voorhees 2004). However, it is also clear that the Q′, AP′ and nDCG′ curves are just as flat as the bpref one.

5 DISCRIMINATIVE POWER

This section compares the robustness of IR metrics to incomplete relevance assessments in terms of discriminative power using Sakai's Bootstrap Sensitivity Method (Sakai 2006b, 2007b). The input to this method are a test collection, a set of runs, an IR metric, and the significance level α for *bootstrap hypothesis tests*(Efron and Tibshirani 1993). Using resampled topic sets, the method conducts a paired bootstrap hypothesis test for every system pair, and computes the discriminative power, i.e., for how many system pairs the IR metric was able to detect a significant difference, and the estimated overall performance difference required to achieve that significance. For this purpose, one thousand bootstrap samples of topics were created for each data set, by sampling with replacement from the original topic set. Details of Sakai's Bootstrap Sensitivity Method can be found elsewhere (Sakai 2006b; 2007b).

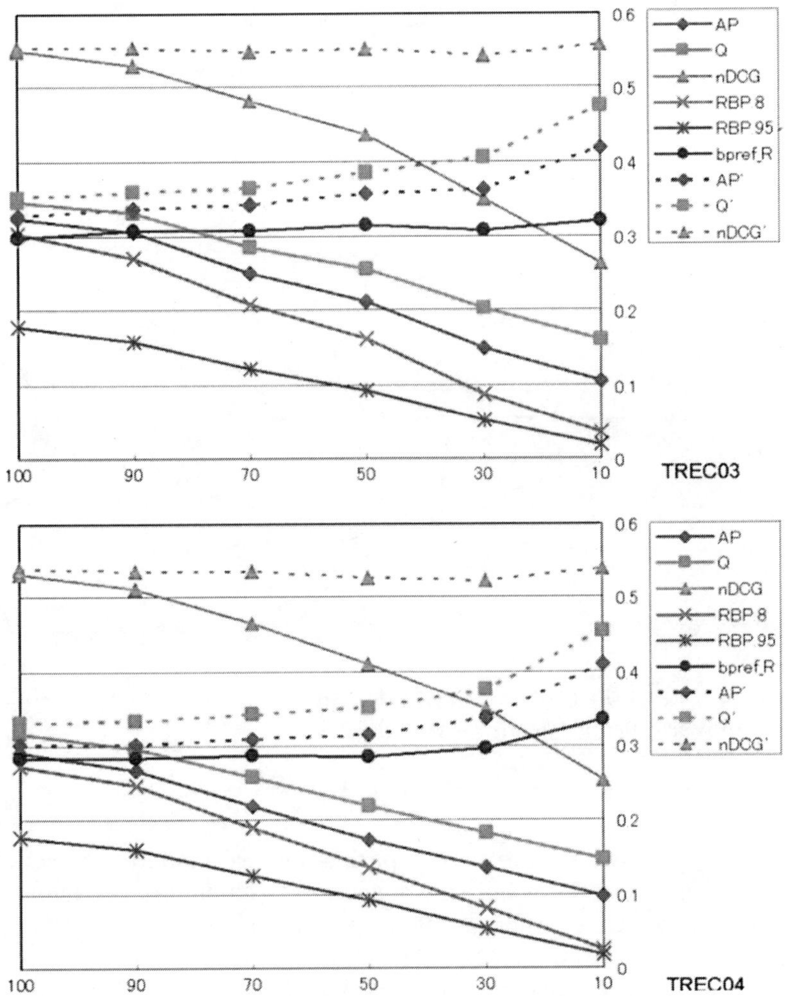

Figure 3. Reduction rate (*x*-axis) versus absolute performance values averaged over 30 runs (*y*-axis) —TREC

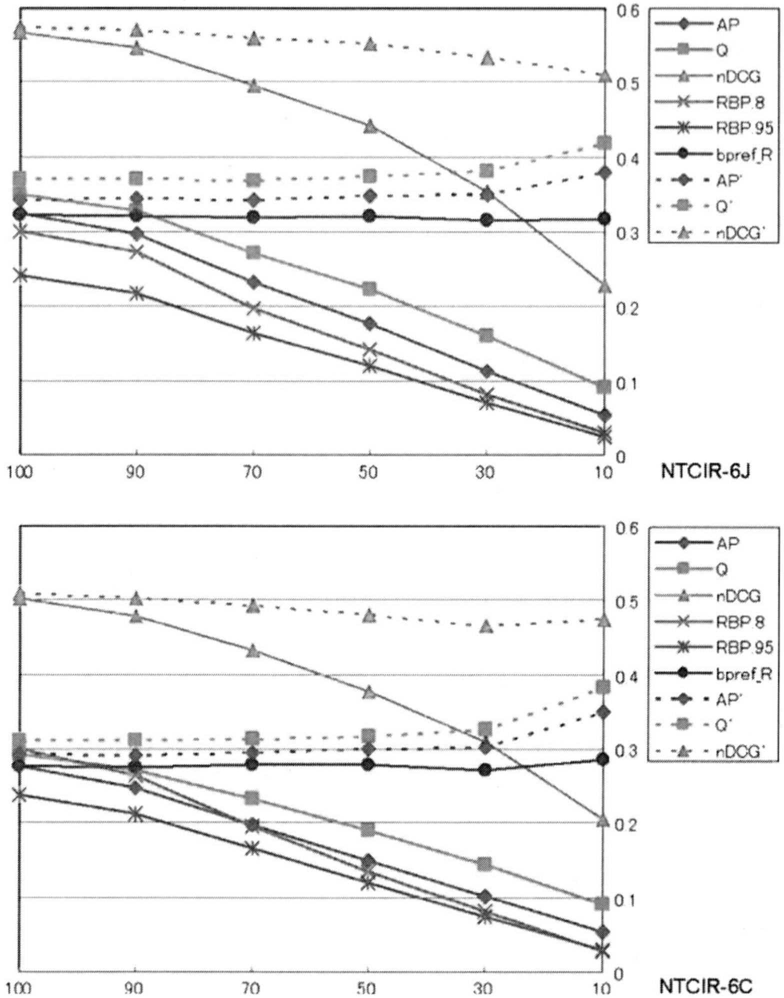

Figure 4. Reduction rate (*x*-axis) versus absolute performance values averaged over 30 runs (*y*-axis) —NTCIR

It should be stressed that we are comparing the discriminative power of different IR metrics while holding α constant in a common experimental environment, where α is the probability of Type I Error, representing the chance of concluding that two systems are different even though the truth is they are equivalent. In other words, IR metrics with high discriminative power are those that can reliably detect meaningful differences between two systems. The Boostrap Sensitivity Method is closely related to the *swap method* proposed by Voorhees and Buckley

(2002) which compares two systems using two different topics sets and examines whether the two outcomes are the same. Unlike the Boostrap Sensitivity Method, the swap method lacks a theoretical foundation and is not directly related to statistical significance tests. However, Sakai (2006b, 2007b) reported that the two methods yield very similar results when used for comparing different IR metrics.

Table 3 compares the discriminative power of Q('), AP('), nDCG('), bpref_R and RBP with the original 100% relevance data. For example, Table 3(a) shows that Q-measure manages to detect a statistical significance for 80 pairs out of 120 (16*15/2) combinations of teams at α = 0.05, and that a difference of around 0.07 is required in order to achieve significance given 50 topics.

Table 3. Discriminative power at α = 0.05 with 100% qrels

	Disc. power (%)	Diff. required		Disc. power (%)	Diff. required
(a) TREC03			(c) NTCIR-6J		
Q	80/120 = 66.7	0.07	nDCG	48/66 = 72.7	0.09
Q′	77/120 = 64.2	0.07	nDCG′	47/66 = 71.2	0.10
AP	77/120 = 64.2	0.07	Q	47/66 = 71.2	0.08
AP′	77/120 = 64.2	0.09	Q′	47/66 = 71.2	0.09
nDCG	71/120 = 59.2	0.08	AP	46/66 = 69.7	0.10
nDCG′	71/120 = 59.2	0.08	AP′	46/66 = 69.7	0.09
bpref_R	69/120 = 57.5	0.08	bpref_R	42/66 = 63.6	0.12
RBP.8	57/120 = 47.5	0.08	RBP.95	42/66 = 63.6	0.07
RBP.95	55/120 = 45.8	0.04	RBP.8	40/66 = 60.6	0.08
RBP.5	45/120 = 37.5	0.12	RBP.5	36/66 = 54.5	0.10
(b) TREC04			(d) NTCIR-6C		
Q	63/91 = 69.2	0.08	nDCG′	43/55 = 78.2	0.10
Q′	62/91 = 68.1	0.08	Q	42/55 = 76.4	0.07
AP	61/91 = 67.0	0.07	nDCG	42/55 = 76.4	0.09
AP′	61/91 = 67.0	0.07	RBP.95	42/55 = 76.4	0.06
nDCG	58/91 = 63.7	0.08	AP′	42/55 = 76.4	0.07
nDCG′	58/91 = 63.7	0.09	bpref_R	42/55 = 76.4	0.08
bpref_R	57/91 = 62.6	0.09	AP	41/55 = 74.5	0.08
RBP.95	45/91 = 49.5	0.05	Q′	40/55 = 72.7	0.08
RBP.8	36/91 = 39.6	0.09	RBP.8	35/55 = 63.6	0.09
RBP.5	30/91 = 33.0	0.12	RBP.5	27/55 = 49.1	0.13

We first summarise Table 3 in words:

- For TREC03 and TREC04, Q(′), AP(′), nDCG(′) and bpref_R are more discriminative than RBP.
- For NTCIR-6J, Q(′), AP(′) and nDCG(′) are more discriminative than bpref_R and RBP.
- For NTCIR-6C, Q(′), AP(′), nDCG(′) bpref_R and RBP.95 are more discriminative than RBP.8 and RBP.5.
- To sum up, the overall winners given 100% relevance data are Q(′), AP(′) and nDCG(′).

It is clear from Table 3 that small values of p for RBP hurt discriminative power. This is probably because a small p makes RBP too top-heavy: as we have seen in Figure 1, using $p = 0.5$ implies that IR systems are more or less evaluated based on the top 10 documents only, which makes evaluation very unreliable (Sakai 2007f). For this reason, we drop RBP.05 from our experiments henceforth.

Figures 5 and 6 show the effect of relevance data reduction on discriminative power for AP(′), Q(′), nDCG(′), bpref_R and RBP.95 and RBP.8. The results are similar to those reported by Sakai (2007a), who used four data sets from NTCIR-3 and NTCIR-5. Table 4 is similar to Table 3 but uses the 10% relevance data, thus representing the "tails" of the curves. We summarise Figs. 5, 6 and Table 4 in words:

- For TREC03 and TREC04, Q′, AP′ and nDCG′ are more robust than other metrics to incomplete relevance assessments. The original nDCG does well for TREC04 but not for TREC03.
- Similarly, for NTCIR-6J and NTCIR-6C, Q′, AP′, nDCG′ and nDCG are the most robust. (Bpref_R appears to do well for NTCIR-6C, but it has a problem, as we shall discuss later using Table 5.)
- RBP.95, AP and RBP.8 are at the bottom of the list, exactly in this order for all four data sets.
- To sum up, the overall winners in terms of robustness to incomplete relevance assessments are Q′, AP′ and nDCG′. AP and RBP clearly lack the robustness. nDCG, Q and bpref_R lie in the middle.

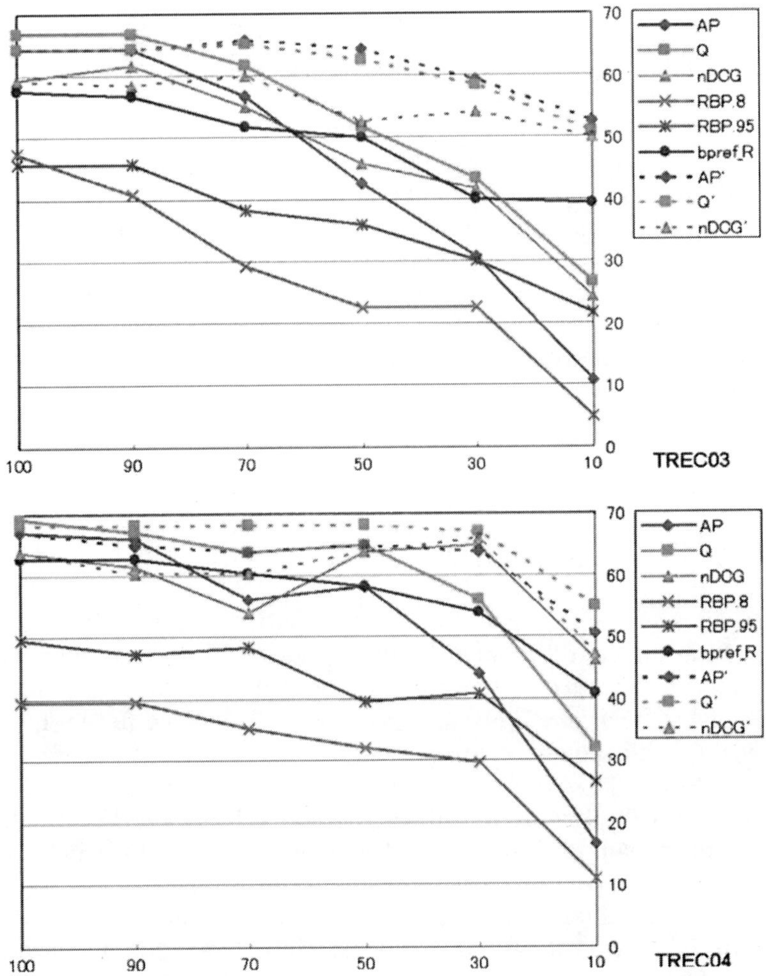

Figure 5. Reduction rate (*x*-axis) versus discriminative power at $\alpha = 0.05$ (*y*-axis)—TREC

The above analysis was based on the number of statistically significant differences detected given incompleteness relevance data. The basic assumption here is that the set of significantly different pairs at *j*% reduction rate is basically a subset of one with the full relevance data. However, it might be the case that most of these conclusions at *j*% reduction rate are in fact inconsistent with the original conclusions with the 100% relevance data. We thus provide an additional analysis in Table 5, which is similar in spirit to the "accuracy" of Bompada et al. (2007). The table compares, for each metric, the set of significantly different pairs at 10% reduction rate with that with the full relevance data. For example, Table 5(a) shows that, for TREC03, AP detected a statistical significance for 13 cases with the 10% relevance data, but two of them (15%) are *not* among the set of cases detected by AP with the 100% relevance data. Assuming that the conclusions with the 100% relevance data are the ground truth, the numbers presented in the table represent "errors". As can be seen, the number of errors are generally small, supporting the aforementioned assumption. Bpref_R, however, appears to be quite unreliable from this viewpoint as well: For example, Table 5(d) shows that as many as 10 cases out of the 39 significant differences detected by bpref_R at 10% reduction rate (See also Table 4(d)) are inconsistent with the original bpref_R results. This, again, is not good news for bpref.

6 RANK CORRELATION

The previous section compared the discriminative power of IR metrics, by focussing on the differences between individual *pairs* of runs. We now compare the*entire* system rankings according to two different IR metrics given the original relevance data, as well as two different qrels, namely, the original one and a reduced one.

Table 6 shows the Kendall's rank correlation values (Voorhees 2001) between each pair of metrics given the original relevance data. As mentioned earlier, we randomly sampled 30 runs from each data set for computing the values: With 30 runs, the correlation is statistically significant if it is over 0.34 (Sakai 2006b); values over 0.9 are shown in bold to indicate high correlations. Note that Kendall' rank correlation is 1 for a pair of identical rankings and −1 if one ranking is a perfect inverse of the other. It can be observed that while the system rankings by AP(′), Q(′), nDCG(′) and bpref_R can be quite similar given the full relevance data, the RBP rankings can be quite different. This alone is not necessarily a flaw: It just means that RBP is measuring something substantially different from the other metrics. Recall that RBP disregards recall.

Figures 7 and 8 show the effect of relevance data reduction on the system ranking for each metric: Thus, the AP ranking at *X*% reduction rate is compared with the original AP ranking, and so on. Table 7 summarises the figures by sorting the metrics by Kendall's rank correlation at 10% reduction rate. Figures 7, 8 and Table 7 show that:

- Q′, AP′ and nDCG′ are consistently among the most robust metrics in terms of system ranking stability. Bpref_R does well for TREC04.
- As Figs. 7 and 8 show, the system rankings by AP and RBP.8 collapse as relevance data is reduced. RBP.95 is also not very good: at 30% reduction rate, its Kendall's rank correlation with the original ranking is as low as that of AP for TREC04 and for NTCIR-6J; it performs as poorly as RBP.8 for NTCIR-6C.

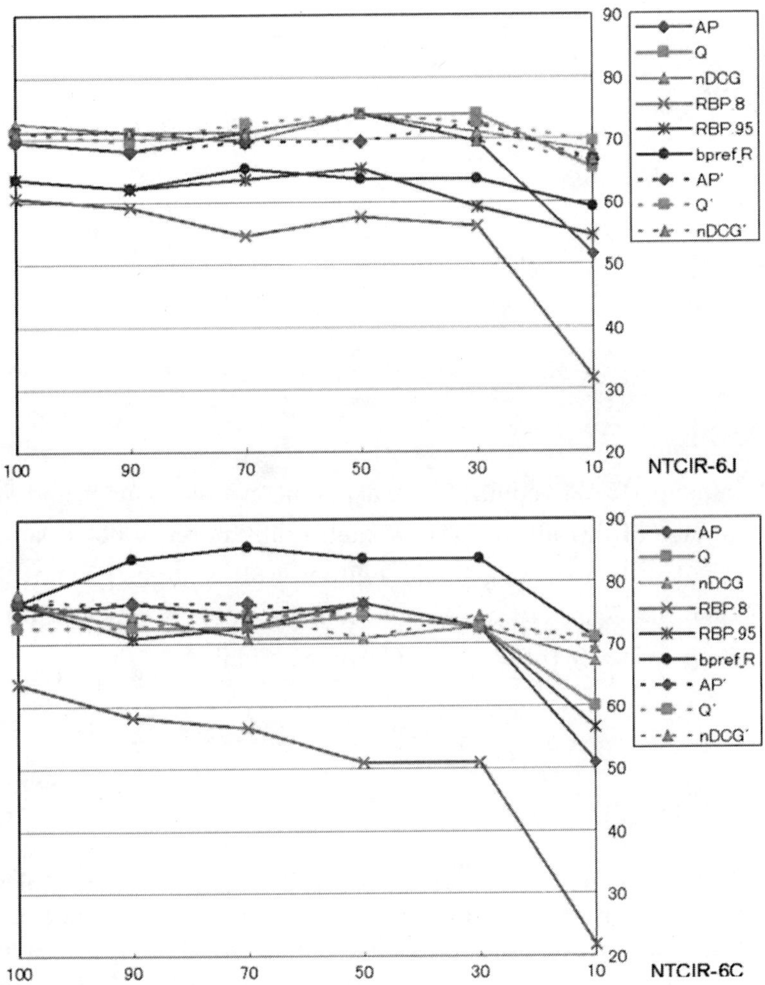

Figure 6. Reduction rate (*x*-axis) versus discriminative power at α = 0.05 (*y*-axis)—NTCIR

Table 4. Discriminative power at α = 0.05 with 10% qrels

	Disc. power (%)	Diff. required		Disc. power (%)	Diff. required
(a) TREC03			(c) NTCIR-6J		
AP′	63/120 = 52.5	0.14	Q′	46/66 = 69.7	0.10
Q′	61/120 = 50.8	0.13	nDCG	45/66 = 68.2	0.06
nDCG′	60/120 = 50.0	0.14	AveP′	44/66 = 66.7	0.11
bpref_R	47/120 = 39.2	0.16	nDCG′	44/66 = 66.7	0.10
Q	32/120 = 26.7	0.09	Q	43/66 = 65.2	0.05
nDCG	29/120 = 24.2	0.09	bpref_R	39/66 = 59.1	0.11
RBP.95	26/120 = 21.7	0.01	RBP.95	36/66 = 54.5	0.01
AP	13/120 = 10.8	0.08	AP	34/66 = 51.5	0.04
RBP.8	6/120 = 5.0	0.03	RBP.8	21/66 = 31.8	0.03
(b) TREC04			(d) NTCIR-6C		
Q′	50/91 = 54.9	0.11	Q′	39/55 = 70.9	0.11
AP′	46/91 = 50.5	0.12	AP′	39/55 = 70.9	0.11
nDCG′	43/91 = 47.3	0.12	bpref_R	39/55 = 70.9	0.12
nDCG	42/91 = 46.2	0.09	nDCG′	38/55 = 69.1	0.12
bpref_R	37/91 = 40.7	0.15	nDCG	37/55 = 67.3	0.06
Q	29/91 = 31.9	0.11	Q	33/55 = 60.0	0.04
RBP.95	24/91 = 26.4	0.01	RBP.95	31/55 = 56.4	0.02
AP	15/91 = 16.5	0.09	AP	28/55 = 50.9	0.04
RBP.8	10/91 = 11.0	0.04	RBP.8	12/55 = 21.8	0.03

- To sum up, Q′, AP′ and nDCG′ are again the overall winners, and the advantage of introducing a new metric like bpref is not clear in terms of system ranking stability either. RBP is not as good as Q′, AP′ and nDCG′ in terms of system ranking stability, even with p = 0.95. Again, nDCG, Q and bpref_R lie in the middle.

7 CONCLUSIONS

This article compared the robustness of IR metrics to incomplete relevance data, using four different sets of graded-relevance test collections with submitted runs—the TREC 2003 and 2004 robust track data and the NTCIR-6 Japanese and Chinese IR data from the crosslingual task. Our discriminative power experiments and rank correlation experiments agreed that Q′, AP′ and nDCG′, the application of Q, AP and nDCG to condensed lists, are more robust than other metrics to relevance data incompleteness; that AP and RBP lack the robustness; and that nDCG, Q and bpref_R lie in the middle. As these results hold across two different evaluation efforts, namely TREC and NTCIR, we believe that these findings are general. It is also interesting that Q′, nDCG′ and AP′ are comparable to one another in terms of robustness to incomplete relevance data, even though Q and nDCG are clearly superior to AP. In other words, the advantage of using graded relevance seems to disappear when condensed lists are used with very incomplete relevance data.

Table 5. Number of significant differences detected with 10% qrels but not with 100% qrels: (i) #significant; (ii) #inconsistent; (iii) percentage

	#Significant	#Inconsistent	%		#Significant	#Inconsistent	%
(a) TREC03				(c) NTCIR-6J			
AP	13	2	15	AP	34	2	6
Q	32	2	6	Q	43	4	9
nDCG	29	0	0	nDCG	45	1	2
RBP.8	6	0	0	RBP.8	21	0	0
RBP.95	26	1	4	RBP.95	36	2	6
bpref_R	47	7	15	bpref_R	39	1	3
AP′	63	3	5	AP′	44	2	5
Q′	61	5	8	Q′	46	1	2
nDCG′	60	5	8	nDCG′	44	0	0
(b) TREC04				(d) NTCIR-6C			
AP	15	1	7	AP	28	1	4
Q	29	0	0	Q	33	0	0
nDCG	42	0	0	nDCG	37	0	0
RBP.8	10	0	0	RBP.8	12	0	0
RBP.95	24	0	0	RBP.95	31	1	3
bpref_R	37	4	11	bpref_R	39	10	26
AP′	46	5	11	AP′	39	1	3
Q′	50	3	6	Q′	39	1	3
nDCG′	43	1	2	nDCG′	38	0	0

Our TREC03, TREC04 and NTCIR-6 results, together with the NTCIR-3 and NTCIR-5 results reported by Sakai (2007a), provide ample evidence that Q′, AP′ and nDCG′ are not only simpler than but also superior to bpref, at least in terms of discriminative power and system ranking stability. Although we have no intention of claiming that Q′, AP′ and nDCG′ are the perfect solution to the problem of relevance data incompleteness, we believe that they are more elegant than introducing metrics like bpref and bpref_N (i.e., RankEff) that lack the "top-heaviness" property of AP by definition.

Table 6. Kendall's rank correlation between different metrics, given 100% qrels

	Q	nDCG	RBP.8	RBP.95	bpref_R	AP′	Q′	nDCG′
(a) TREC03								
AP	**.931**	.857	.706	.848	**.922**	**.982**	**.931**	.867
Q	—	.844	.655	.807	.871	**.949**	**.991**	.853
nDCG	—	—	.775	.853	.844	.857	.844	**.991**
RBP.8	—	—	—	.821	.747	.697	.655	.775
RBP.95	—	—	—	—	.899	.839	.798	.853
bpref_R	—	—	—	—	—	**.913**	.862	.844
AP′	—	—	—	—	—	—	**.949**	.867
Q′	—	—	—	—	—	—	—	.853
(b) TREC04								
AP	**.968**	**.940**	.747	.890	**.968**	**.977**	**.945**	**.945**
Q	—	**.936**	.733	.876	**.954**	**.972**	**.977**	**.940**
nDCG	—	—	.770	**.903**	**.936**	**.936**	**.922**	**.977**
RBP.8	—	—	—	.821	.770	.733	.710	.756
RBP.95	—	—	—	—	**.913**	.876	.853	.890
bpref_R	—	—	—	—	—	**.945**	**.931**	**.931**
AP′	—	—	—	—	—	—	**.959**	**.940**
Q′	—	—	—	—	—	—	—	**.945**
(c) NTCIR-6J								
AP	**.968**	.834	.770	.857	**.908**	**.936**	**.903**	.811
Q	—	.857	.793	.880	**.913**	**.940**	**.936**	.834
nDCG	—	—	.862	.885	.834	.853	.894	**.968**
RBP.8	—	—	—	.830	.770	.789	.802	.839
RBP.95	—	—	—	—	.867	.885	.890	.853
bpref_R	—	—	—	—	—	**.972**	**.922**	.830
AP′	—	—	—	—	—	—	**.940**	.848
Q′	—	—	—	—	—	—	—	.880
(d) NTCIR-6C								
AP	**.963**	**.903**	.885	**.949**	**.972**	**.986**	**.936**	**.908**
Q	—	**.922**	.857	**.922**	**.936**	**.949**	**.972**	**.926**
nDCG	—	—	.880	**.945**	**.913**	.899	**.949**	**.995**
RBP.8	—	—	—	**.926**	**.903**	.880	.839	.876
RBP.95	—	—	—	—	**.968**	**.945**	**.913**	**.940**
bpref_R	—	—	—	—	—	**.977**	**.917**	**.908**
AP′	—	—	—	—	—	—	**.922**	**.903**
Q′	—	—	—	—	—	—	—	**.954**

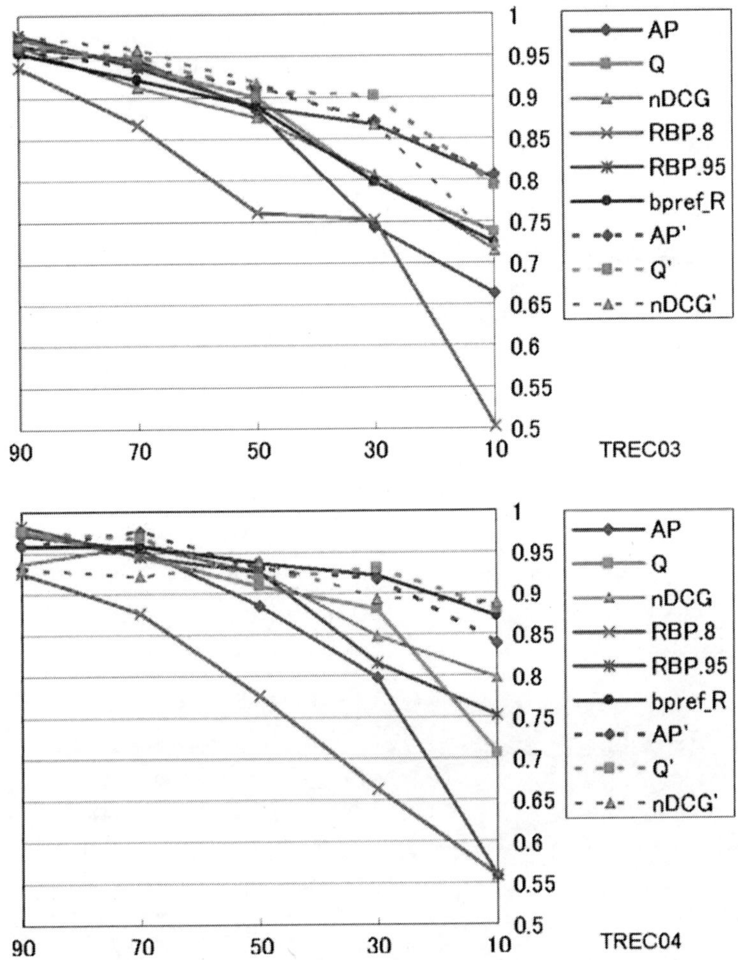

Figure 7. Reduction rate (*x*-axis) versus Kendall's rank correlation with the 100%-qrels ranking (*y*-axis) —TREC

Even though Moffat et al. (2007) claimed that RBP is suitable for evaluation with incomplete relevance data as its error due to unjudged documents can be quantified, we demonstrated that it has weaknesses. While RBP is interesting in that it is independent of recall, because of this very feature, it often does not equal one even for an ideal ranked output. For example, as we have discussed using Table 1, an ideal output for a topic with 10 (regular) relevant documents may receive an RBP of .4013, while an ideal output for a topic with 100 (regular) relevant documents may receive an RBP of .9941. Whether it is good to average such a measurement across topics is debatable. Moreover, our experimental results showed that small values of p make RBP unreliable, and that RBP is not as robust to incomplete relevance data as Q′, AP′ and nDCG′ in terms of discriminative power and system ranking stability, even with $p = 0.95$.

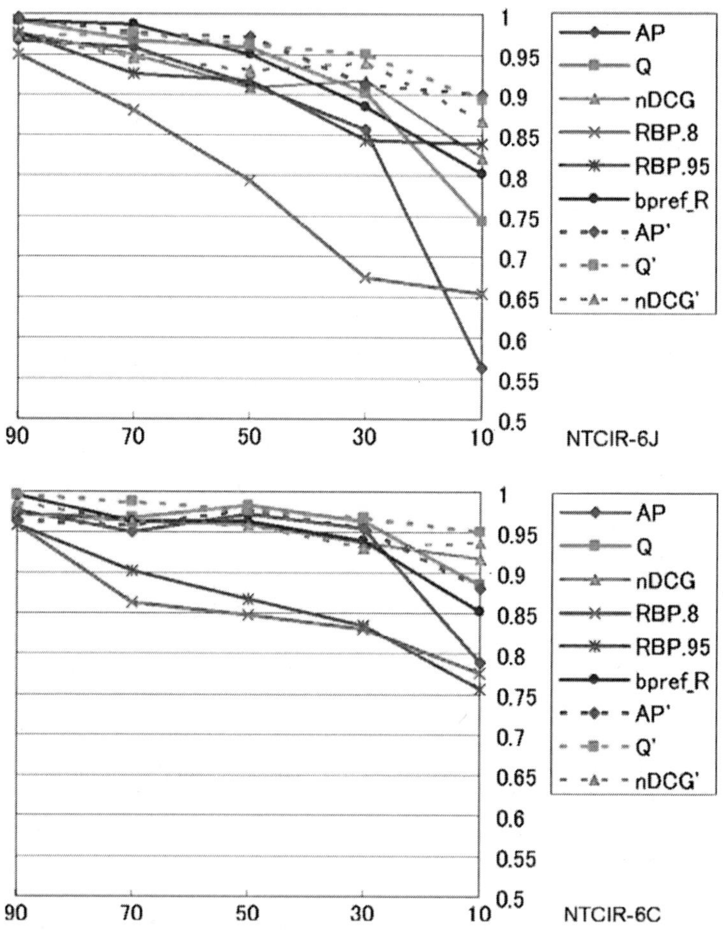

Figure 8. Reduction rate (x-axis) versus Kendall's rank correlation with the 100%-qrels ranking (y-axis)—NTCIR

Table 7. Kendall's rank correlation: 100% vs 10% qrels for each metric

(a) TREC03		(c) NTCIR-6J	
AP′	.807	AP′	.899
RBP.95	.802	Q′	.894
Q′	.793	nDCG′	.867
Q	.738	RBP.95	.839
nDCG′	.724	nDCG	.821
bpref_R	.724	bpref_R	.802
nDCG	.715	Q	.743
AP	.664	RBP.8	.655
RBP.8	.503	AP	.563
(b) TREC04		(d) NTCIR-6C	
nDCG′	.890	Q′	.949
Q′	.880	nDCG′	.936
bpref_R	.871	nDCG	.917
AP′	.839	Q	.885
nDCG	.798	AP′	.880
RBP.95	.752	bpref_R	.853
Q	.706	AP	.789
AP	.559	RBP.8	.775
RBP.8	.559	RBP.95	.756

The fact that Q′, AP′ and nDCG′ perform clearly and consistently better than the original Q, AP and nDCG in an incomplete relevance environment implies the following: The assumption that all unjudged documents are nonrelevant is not good; It is much better to treat all unjudged documents as if they never existed, in order to let judged relevant and judged nonrelevant documents move up the ranks and hence serve as stronger pieces of evidence for computing system effectiveness.

It should be recalled, however, that we used random sampling from the original qrels in order to artificially create very incomplete test collections. We shall discuss the effect of using shallow pools and that of using fewer participating teams for forming relevance assessments elsewhere Sakai (2008a, 2008b). Moreover, although we examined the IR metrics in terms of discriminative power and Kendall's rank correlation, there may be other criteria for choosing "good" metrics. "Simplicity" and "intuitiveness" are but a few examples, although they are difficult to quantify. Establishing a standard set of criteria for metric selection is an important goal of our future research.

As we mentioned in Sect. 2, our present study takes the approach of choosing IR metrics *given* a test collection with incomplete relevance data. However, the approach of constructing reliable test collections efficiently (e.g., work by Carterette et al. 2006) is equally important, and combining these two approaches is probably even more so. That is, IR metrics should perhaps be designed by taking the process of test collection construction into account. This is another research topic that needs to be explored.

FOOTNOTES

[1]http://trec.nist.gov/

[2]http://ntcir.nii.ac.jp/

[3]http://www.clef-campaign.org/

[4]Subcollection AP requires even more knowledge, namely, how small the subcollection with relevance assessments is compared to the entire document collection.

[5]AP, on the other hand, has a different weakness, in that it can be one for a suboptimal ranked output in a graded relevance environment. To be more specific, AP is one as long as all top *R* documents are at least somewhat relevant: It does not matter if partially relevant documents are retrieved above highly relevant ones.

[6]One way to avoid this problem is to discount the gains by loga(r+1) for every rank *r* (Burges et al. 2005). However, this cancels out the logarithm base *a*, which was originally intended as a parameter for reflecting the user's patience (Järvelin and Kekäläinen 2002).

[7]The statistical significance of Kendall's rank correlation depends directly on the number of runs (Sakai 2006b).

ACKNOWLEDGMENTS

We thank William Webber, Justin Zobel and Alistair Moffat for their criticisms and constructive comments. We also thank Ellen Voorhees for letting us use the TREC robust track data, and Alistair Moffat and Justin Zobel for providing their unpublished manuscript (Moffat and Zobel 2008).

REFERENCES

1. Aslam, J. A., & Savell, R. (2003). On the effectiveness of evaluating retrieval systems in the absence of relevance judgments. In *ACM SIGIR 2003 Proceedings* (pp. 361–362).

2. Aslam, J. A., Pavlu, V., & Yilmaz, E. (2006). A statistical method for system evaluation using incomplete judgments. In *ACM SIGIR 2006 Proceedings* (pp. 541–548).

3. Bompada, T., Chang, C.-C., Chen, J., Kumar, R., & Shenoy, R. (2007). On the robustness of relevance measures with incomplete judgments. In *ACM SIGIR 2007 Proceedings* (pp. 359–366).

4. Buckley, C., & Voorhees, E. M. (2004). Retrieval evaluation with incomplete information. In *ACM SIGIR 2004 Proceedings* (pp. 25–32).

5. Burges, C., Shaked, T., Renshaw, E., Lazier, A., Deeds, M., Hamilton, N., & Hullender, G. (2005). Learning to rank using gradient descent. In *ACM ICML 2005 Proceedings* (pp. 89–96).

6. Büttcher, S., Clarke, C. L. A., Yeung, P. C. K., & Soboroff, I. (2007). Reliable information retrieval evaluation with incomplete and biased judgements. In *ACM SIGIR 2007 Proceedings* (pp. 63–70).

7. Carterette, B., Allan, J., & Sitaraman, R. (2006). Minimal test collections for retrieval evaluation. In *ACM SIGIR 2006 Proceedings* (pp. 268–275).

8. Cleverdon, C. W. (1967). The cranfield tests on index language devices. In *Aslib Proceedings* (Vol. 19, pp. 173–192).

9. Cormack, G. V., Palmer, C. R., & Clarke, C. L. A. (1998). Efficient construction of large test collections. In *ACM SIGIR '98 Proceedings* (pp. 282–289).

10. De Beer, J., & Moens, M.-F. (2006). Rpref—a generalization of bpref towards graded relevance judgments. In*ACM SIGIR 2006 Proceedings* (pp. 637–638).

11. Efron, B., & Tibshirani, R. (1993). *Introduction to the bootstrap.* Chapman & Hall/CRC.

12. Grönqvist, L. (2005). Evaluating latent semantic vector models with synonym tests and document retrieval. In*ELECTRA Workshop—Methodologies and Evaluating of Lexical Cohesion Techniques in Real-World Applications*(pp. 86–88).

13. Järvelin, K., & Kekäläinen, J. (2002). Cumulated gain-based evaluation of IR techniques. *ACM Transactions on Information Systems, 20*(4), 422–446.

14. Kando, N. (2007). *Overview of the Sixth NTCIR Workshop.* In *NTCIR-6 Proceedings* (pp. i–ix).

15. Kekäläinen, J. (2005). Binary and graded relevance in IR evaluations—comparison of the effects on ranking of IR systems. *Information Processing and Management, 41*, 1019–1033.

16. Moffat, A., Webber, W., & Zobel, J. (2007). Strategic system comparisons via targeted relevance judgments. In*ACM SIGIR 2007 Proceedings* (pp. 375–382).

17. Moffat, A., & Zobel, J. (2008). Rank-biased precision for measurement of retrieval effectiveness (under review).

18. Sakai, T. (2006a). Bootstrap-based comparisons of IR metrics for finding one relevant document. In *AIRS 2006: Lecture Notes in Computer Science 4182* (pp. 374–389). Springer-Verlag.

19. Sakai, T. (2006b). Evaluating evaluation metrics based on the bootstrap. In *ACM SIGIR 2006 Proceedings* (pp. 525–532).

20. Sakai, T. (2006c). On the task of finding one highly relevant document with high precision. *Information Processing Society of Japan Transactions on Databases, 47*(SIG4(TOD29)), 13–27

21. Sakai, T. (2007a). Alternatives to Bpref. In *ACM SIGIR 2007 Proceedings* (pp. 71–78).

22. Sakai, T. (2007b). Evaluating information retrieval metrics based on bootstrap hypothesis tests. *Information Processing Society of Japan Transactions on Databases 48*(SIG9(TOD35)), 11–28

23. Sakai, T. (2007c). evaluating the task of finding one relevant document using incomplete relevance data. In *Forum on Information Technology 2007 Information Technology Letters* (pp. 91–94).

24. Sakai, T. (2007d). On penalising late arrival of relevant documents in information retrieval evaluation with graded relevance. In *Proceedings of the First International Workshop on Evaluating Information Acess (EVIA 2007)* (pp. 32–43).

25. Sakai, T. (2007e). On the properties of evaluation metrics for finding one highly relevant document. *Information Processing Society of Japan Transactions on Databases, 48*(SIG9(TOD35)), 29–46.

26. Sakai, T. (2007f). On the reliability of information retrieval metrics based on graded relevance. *Information Processing and Management, 43*(2), 531–548.

27. Sakai, T. (2008a). Comparing metrics across TREC and NTCIR: The robustness to pool depth bias (under review).

28. Sakai, T. (2008b). Comparing metrics across TREC and NTCIR: The robustness to system bias (under review).

29. Sakai, T., & Sparck Jones, K. (2001). Generic summaries for indexing in information retrieval. In *ACM SIGIR 2001 Proceedings* (pp. 190–198).

30. Soboroff, I., Nicholas, C., & Cahan, P. (2001). Ranking retrieval systems without relevance judgments. In *ACM SIGIR 2001 Proceedings* (pp. 66–73).

31. Sormunen, E. (2002). Liberal relevance criteria of TREC—counting on negligible documents? In *ACM SIGIR 2002 Proceedings* (pp. 324–330).

32. Voorhees, E. M. (2001). Evaluation by highly relevant documents. In *ACM SIGIR 2001 Proceedings* (pp. 74–82).

33. Voorhees, E. M. (2002). The effect of topic set size on retrieval experiment error. In *ACM SIGIR 2002 Proceedings* (pp. 316–323).

34. Voorhees, E. M. (2004). Overview of the TREC 2003 robust retrieval Track. In *TREC 2003 Proceedings.*

35. Voorhees, E. M. (2005). Overview of the TREC 2004 robust retrieval track. In *TREC 2004 proceedings.*

36. Voorhees, E. M., & Buckley, C. (2002). The philosophy of information retrieval evaluation. In *Proceedings of CLEF 2001, Lecture Notes in Computer Science 2406* (pp. 355–370).

37. Yilmaz, E., & Aslam, J. A. (2006). Estimating average precision with incomplete and imperfect judgments. In*ACM CIKM 2006 Proceedings* (pp. 102–111).

38. Zobel, J. (1998). How Reliable are the results of large-scale information retrieval experiments? In *ACM SIGIR '98 Proceedings* (pp. 307–314).

4

MODELING ACTIONS OF PUBMED USERS WITH *N*-GRAM LANGUAGE MODELS

Jimmy Lin[1], W. John Wilbur[2]

[1]The iSchool, College of Information Studies, University of Maryland, College Park, MD, USA

[2]National Center for Biotechnology Information, National Library of Medicine, Bethesda, MD, USA

ABSTRACT

Transaction logs from online search engines are valuable for two reasons: First, they provide insight into human information-seeking behavior. Second, log data can be used to train user models, which can then be applied to improve retrieval systems. This article presents a study of logs from PubMed®, the public gateway to the MEDLINE® database of bibliographic records from the medical and biomedical primary literature. Unlike most previous studies on general Web search, our work examines user activities with a highly-specialized search engine. We encode user actions as string sequences and model these sequences using *n*-gram language models. The models are evaluated in terms of perplexity and in a sequence prediction task. They help us better understand how PubMed users search for information and provide an enabler for improving users' search experience.

KEYWORDS

Search behavior, Query log analysis

1 INTRODUCTION

Information seeking is fundamentally an iterative activity that involves multiple interactions between a user and a search system. With the advent of online search engines, it has become standard practice to store records of users' activities—commonly known as search transaction logs (or query logs, as a special case). These resources are useful for two main reasons: First, they shed light on human information-seeking behavior—what users want and how they go about accomplishing it (e.g., Silverstein et al. 1999; Beitzel et al. 2004; Rose and Levinson 2004; Jansen and Spink 2006). Second, log data can be exploited to improve the user's search experience; examples include query suggestion (Anick2003; Cui et al. 2003) and improved ranking (Agichtein et al. 2006; Joachims et al.2007).

The nature of users' interactions with search engines depends on many factors, including characteristics of the user, type of task, problem domain, etc. While there has been a significant amount of work on analysis of search logs, previous studies have almost exclusively focused on general Web search; see Jansen and Spink (2006) for an overview. In this article, we present an analysis of logs from the PubMed search engine (more details in Sect. 2). Note that due to the subject domain of PubMed (the life sciences) and the demographic profile of its users (biologists, physicians, clinical researchers, etc.), our dataset differs significantly from typical Web search logs, such as those collected by Google or Yahoo, which represent a broader demographic and a greater variety of user needs. However, these differences are not the subject of this article.

Our work has two main goals: First, we wish to better understand how users interact with PubMed (Sects. 3, 5). Second, we wish to build computational models of user–system interactions as a first step to improving the search experience. This article explores the idea that users' actions can be encoded as a string sequence and modeled using *n*-gram language models (Sect. 4). We evaluate these models both in terms of perplexity and in a sequence prediction task.

2 PUBMED TRANSACTION LOGS

PubMed is the public gateway to MEDLINE, the authoritative repository of bibliographic records from the medical and biomedical primary literature, and more broadly, topics in the life sciences ranging from biophysics to public health.1 Both resources are maintained by the U.S. National Library of Medicine (NLM). As of March 2008, MEDLINE contains over 17.8 m records (called citations) dating back to 1949, all with basic bibliographic information. In addition, most records (especially those added in more recent decades) contain abstract text, and increasingly, links to full text. Citations are added to MEDLINE at a rate of approximately 65k records a month. PubMed is frequently used by physicians, scientists (e.g., biologists, biochemists), and lay-people to solve complex tasks (particularly the first two groups). Although PubMed provides access to a wealth of resources (e.g., gene databases, sequence information, etc.), this study focuses specifically on users' interaction with MEDLINE.

PubMed is a sophisticated Boolean search engine that allows users to query not only on title and abstract text, but also on metadata fields (e.g., journal or author) and using controlled vocabulary MeSH® terms. PubMed also provides a number of pre-defined "search templates", which allow users to narrow the scope of the articles that are retrieved (Haynes et al. 1994). These filters are implemented as fixed Boolean query fragments that the system automatically appends to each query.

This work examines transaction logs gathered over an 8-day span, June 20–27, 2007. The basic unit of analysis is the session, which is tracked through a browser cookie. Sessions are comprised of transactions, each of which corresponds to a CGI invocation. Due to the nature of this tracking mechanism, a user who engages PubMed with multiple browser windows (or tabs) will show up as a single session, since there is no effective way to separate the source of the CGI requests. Note that our definition of a session is very coarse-grained; we explore different segmentation techniques in Sect. 3.

The logs contain a wealth of information, including timestamp and details of the CGI invocation (encoded parameters), which allows us to reconstruct with reasonable accuracy the actions of a particular user. Certain client-side actions, such as use of the browser "back" button, are not captured, although it is possible to infer some of these behaviors. This data provides a unique glimpse into the actions of PubMed users—although PubMed queries have been previously studied (Herskovic et al. 2007), this work represents the first systematic analysis of session data.

Although there have been previous studies of transaction logs from search systems for the medical literature, e.g., studies on PaperChase (Horowitz et al.1983; King 1991) and Grateful Med (Cahan 1989), our work involves significantly more data and users. It is our intention to make an appropriately-processed version of this dataset available to the research community.

In addition to the domain, we note another unique characteristic of our logs. Whereas most datasets studied by academic researchers contain only search queries (Jansen and Spink 2006), our logs capture all interactions between the user and PubMed. In fact, search queries account for only 33% of user actions. A more complete record allows us to construct richer models of users' activities and understand their browsing behaviors (see Sect. 5).

Over the 8-day span, we observed a total of 8.68 m sessions, totaling 41.8 m transactions (for clarity, we refer to this as the raw dataset). A number of filters were first applied to discard sessions not of interest. We found that 63.2%, or 5.49 m sessions, consisted of only one transaction—most of these represent direct access into MEDLINE, e.g., through an embedded link or another search engine; this number is consistent with previous studies on Web search (Jansen and Spink2006). Although single-transaction sessions account for a large portion of all traffic, we discarded them since they do not represent interesting user behavior. At the other end of the spectrum, we discarded all sessions with more than 500 transactions (an arbitrary threshold), since they were likely to be robots spidering the collection. This removes only 162 sessions, but 271k transactions. Finally, we removed sessions in which the user was not primarily engaged with MEDLINE. In addition to bibliographic records, PubMed provides access to other databases (e.g., gene records), which are not of interest for this study. As a heuristic, we discarded sessions in which more than half the transactions did not involve MEDLINE; this removed an additional 2.72 m sessions. Ultimately, we ended up with a dataset containing 476k sessions, 7.65 m transactions; for clarity, we refer to this as the filtered dataset. Although the size of the dataset after applying these filters is a fraction of the original raw dataset, we argue that the processing steps allow us to focus on "interesting" user behavior, at least for the purposes of this study. Note that similar filtering is often applied to general Web query logs so that a researcher can focus on phenomena of interest. For example, previous analyses (Rose and Levinson 2004; Broder 2002) have found that, depending on the actual dataset, up to a quarter of all queries are navigational in nature, and up to a third of all queries are intended to locate a particular resource (e.g., finding a currency converter or downloading a video clip) rather than to gather information.

Building on Murray et al. (2006), the main idea explored in this work is to analyze user activities with *n*-gram language models; cf. (Chen and Cooper 2002). This is accomplished by first mapping classes of user actions into symbols drawn from a finite alphabet. Our encoding is shown in Table 1. Thus, sessions can be represented as sequences of symbols, and building computational models of users' activities can be recast into a sequence modeling problem. For reference, Table 1also shows the distribution of the 7.65 m user actions in the filtered dataset. The following are the beginnings of three sample sessions encoded according to our scheme:

QNRRRRLRQNRQQQQQ...
QNQQQQQQNQNQQQN...
QNNNNNQNRQVNRRQNR...

Table 1. Mapping of PubMed transactions into symbols representing different action types

Query: the user issued a PubMed query	2518955	32.9%
Retrieve: the user clicked on a search result to display a MEDLINE record, which contains bibliographic information and abstract text (in most cases). This view provides a link to the full-text article if available, but these actions are not recorded in our logs. The view of the MEDLINE record also contains links to five related articles (see below)	3036158	39.7%
Next: the user requested the next page of search results. PubMed displays 20 hits per page	658214	8.6%
Related Link: the user clicked on a related article link. Whenever the user examines a MEDLINE record, the right panel of the browser window is automatically populated with titles of the 5 related articles most similar in content to the one currently being examined (Lin and Wilbur 2007). This feature allows the user to browse citations without explicitly issuing new queries (Lin et al. 2008)	284974	3.7%
More links: the user requested more related articles. In the MEDLINE record view, only the top 5 related articles are shown; however, more related articles can be requested via a separate link	52546	0.7%
Modify View: the user manipulated search results using advanced features in the "Limits" tab of the PubMed interface. Options in that tab allow the user to restrict the retrieved set in terms of journal, author, availability, date, etc.	515720	6.7%
P (other MEDLINE): catch-all category for other MEDLINE-related actions. These include actions in the "Preview/Index", "History", "Clipboard", and "Details" tabs of the PubMed interface, as well as clicks on the "LinkOut" resource supplied by external providers	287604	3.8%
X (other PubMed): catch-all category for other actions not involving MEDLINE (e.g., access to gene records)	291104	3.8%
Total	7645275	

Even disregarding details such as the actual query and the timestamp, it is possible to construct an understanding of user behavior in each of these cases. In the first example, the sequence of R's indicate that the user examined four MEDLINE citations on the same result page. The long sequences of Q's in the second session suggest that the user had difficulty formulating a good PubMed query. The sequence of N's in the third example indicates that the user looked at several pages of results without selecting a MEDLINE citation to examine in detail, before finally giving up and issuing a different query.

3 SESSION SEGMENTATION

We first tackled the session segmentation problem. Note that there is no consistent definition of a "session" in the literature. For some researchers, sessions are semantic in nature. As an example, He and Göker (2000) proposed that "the start and end of a session are the points where the role behind a query changes", which intimately ties the notion of a session to information-seeking tasks. For others, sessions are non-semantic constructs, defined purely in terms of temporal inactivity (Silverstein et al. 1999) or in terms of the granularity at which the log data is gathered (Jansen et al. 2000). We adopted the second approach, defining sessions in terms of browser cookies. Despite differences in terminology, there is nevertheless consensus that user activities can be segmented into "meaningful" units.

To reduce confusion, we use the term "episode" to refer to meaningful units of activities within sessions. An episode might, for example, correspond to a specific information-seeking task. Although the identification of these units is difficult without understanding the semantics of users' queries, temporal information alone provides significant information (Catledge and Pitkow 1995; He and Goker 2000). That is, a system could segment a sequence of user actions into episodes based on periods of inactivity. This method is extremely fast (a prerequisite for online processing) and sidesteps thorny issues of query semantics. In addition, methods based purely on temporal information do not raise privacy concerns, compared to methods that depend on the actual semantics of the user action (e.g., queries and clickthroughs).

What should this inactivity threshold be? Naturally, we face a tradeoff: the longer the duration, the more likely we are to create episodes that span multiple, possibly unrelated, information-seeking tasks. On the other hand, shorter durations may yield incomplete episode fragments. For example, if the user spends a long time reading a result page, and then returns to continue browsing additional search results, the system might infer multiple episodes if the threshold is too short.

However, this tradeoff can be used to our advantage: in absence of "ground truth", segmenting sessions into episodes based on different thresholds can serve as a probe into users' behavior, by analyzing the characteristics of the resulting episodes. We did exactly this, segmenting sessions based on different inactivity thresholds, ranging from 5 to 60 min in 5 min increments. The line marked by diamonds in Figure 1 denotes the total number of episodes that is generated at each threshold. Naturally, smaller thresholds generate more episodes. The line marked by filled squares represents the number of singleton episodes (i.e., episodes with only one transaction); the line marked by empty squares represents the number of singleton episodes consisting of a citation retrieval. The ratio between these two values is expressed as a percentage above the filled squares, i.e., the percentage of singleton episodes where the user retrieved a MEDLINE record. Note that from the citation view, users can access available full text (these actions are not recorded in our logs). We see that nearly 70% of all singleton episodes are retrievals.

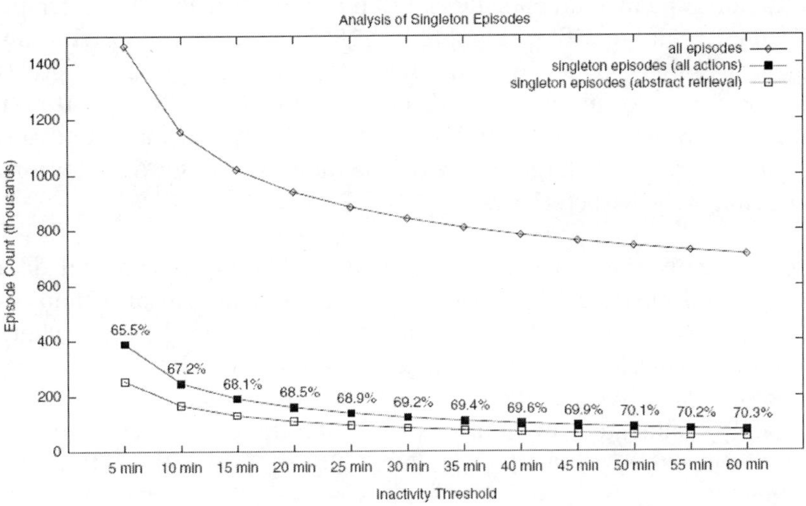

Figure 1. Characteristics of episodes generated by applying different thresholds to segment sessions. Values above each filled square indicate the percentage of singleton episodes that consists of a single retrieve action

Combined with the observation that about 60% of episodes end with a retrieval (regardless of threshold), we infer that singleton retrieval episodes are likely to be an extension of the previous episode—since the user is directly accessing MEDLINE content. By comparing these results with figures cited in Jansen and Spink (2004), we conclude that PubMed users spend a longer period of time examining content. This makes sense given the more complex tasks that PubMed users typically engage in—for example, physicians searching for clinical evidence in the context of patient care (De Groote and Dorsch 2003; Herskovic et al. 2007), biologists combing the literature for studies that link a particular gene to a disease (Hersh et al. 2005, 2007).

Figure 2 shows the distribution of episodes in terms of two different measures of length: number of transactions (top) and duration (bottom). For both graphs, we show the results of segmentation based on three representative thresholds: 5, 15, and 30 min. The plots for even longer thresholds are not substantially different from the plot for the 30-min threshold, and unfortunately showing any more lines would result in too much clutter. Duration is measured as the time difference between the first transaction of the episode and the last transaction (thus singleton episodes have zero duration). The bottom graph shows duration binned in increments of 5 min, e.g., increment 6 corresponds to all durations between 25 and 30 min long. In terms of number of interactions, the median episode length is 3 with an inactivity threshold of 5 and 10 min; 4 with a threshold of 15, 20, 25, 30 min; 5 with any longer thresholds. In terms of duration, the median episode length is less than 5 min for an inactivity threshold of 5, 10, 15, 20, 25, 30 min, rising to between 5 and 10 min for any longer thresholds. In both cases, the means are much larger than the medians since the distributions exhibit long tail characteristics.

Note that a meaningful comparison between PubMed data and data from general-purpose Web search engines (e.g., Google or Yahoo) is difficult, since most existing logs available to academic researchers contain only queries; see overview in Jansen and Spink (2006). In the case of PubMed, we have shown that episode boundaries cannot be accurately delineated without records of citation retrievals, since PubMed users may spend significant time examining MEDLINE citations.

These results seem to suggest that the choice of threshold is perhaps not as significant as one might think. With the exception of the 5-min threshold (which seems too short given the discussion above), the plots for episode length distribution don't actually differ by much. In terms of transactions, longer thresholds are primarily reducing the number of singleton episodes, which are mostly citation retrievals. This has the effect of appending an additional 'R' symbol at the end of other sequences, and has a relatively minor effect for the experiments we describe in subsequent sections. Similarly, different thresholds have relatively minor impact on episode duration (disregarding the 5-min threshold).

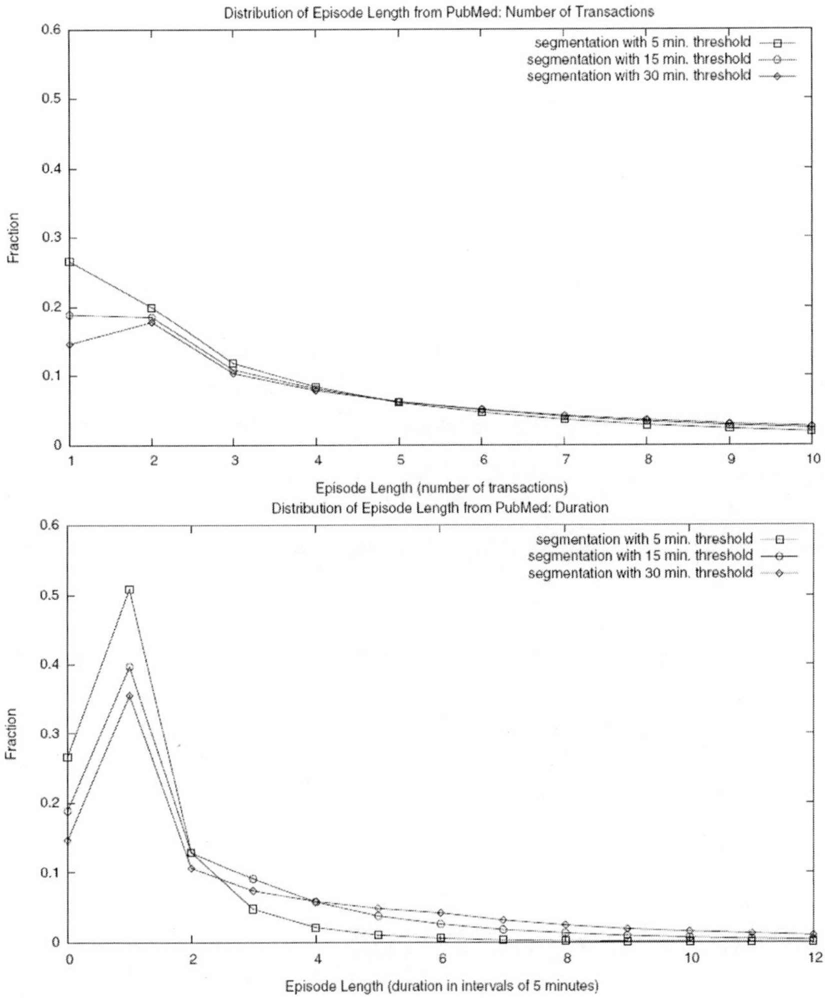

Figure 2. Distribution of episode length, in terms of number of transactions (top) and duration (bottom). Duration is binned in 5 min intervals (e.g., '5' represents intervals between 20 and 25 min)

To determine if there is any value in our simple notion of episodes and to better facilitate subsequent computational modeling, we prepared another dataset, which we refer to as the episode dataset. From the original set of filtered sessions we obtained a set of episodes by first segmenting the logs using a 30-min threshold. We then discarded the following:

- all singleton episodes (since they cannot contribute to the user action prediction task we describe in Sect. 4);
- all episodes that do not involve MEDLINE, i.e., consisting exclusively of the symbol 'X' (since our study focuses on searching MEDLINE);
- all episodes that do not begin with a query (since they often represent direct access into MEDLINE, i.e., through another search engine, and we are primarily interested in studying search *with* PubMed).

From the 476k sessions in the filtered dataset, we ended up with 373k episodes containing 4.34 m transactions. For convenience, a summary of all datasets mentioned in this article is shown in Table 2.

Table 2. Summary of the three datasets referenced in this article

Dataset	Size	Brief description
Raw	8.68 m sessions, 41.8 m trans.	Unprocessed logs
Filtered	476k sessions, 7.65 m trans.	Filtered sessions
Episode	373k episodes, 4.34 m trans.	Filtered episodes

4 MODELING USER ACTIVITIES

One advantage of encoding user actions as sequences of symbols is the ability to use standard natural language processing techniques to build computational models of user activity. In this work, we experimented with *n*-gram language models. Language models define a probability distribution over string sequences:

$$P(w_1 w_2 w_3 \ldots w_{n-1} w_n) \equiv P(w_1^n) \quad (1)$$

In language processing, this typically means sequences of words; see Manning and Schütze (1999) for an overview. In our application, we are modeling sequences of symbols, each of which represent a user action. By the chain rule of probability theory:

$$P(w_1^n) = P(w_1)P(w_2|w_1)P(w_3|w_1 w_2)\ldots P(w_n|w_1 w_2 \ldots w_{n-1}) \quad (2)$$

$$= P(w_1)P(w_2|w_1)P(w_3|w_1^2)\ldots P(w_n|w_1^{n-1}) \quad (3)$$

$$= \prod_{k=1}^{n} P(w_k|w_1^{k-1}) \quad (4)$$

Due to the extremely large number of parameters involved in estimating such a model, it is customary to make the *Markov assumption*, that the sequence histories only depend on prior local context. That is, an n-gram language model is equivalent to a $(n - 1)$-order Markov model. Thus, we can approximate $P(w_k|w_1^{k-1})$ as follows:

$$\text{bigrams: } P(w_k|w_1^{k-1}) \approx P(w_k|w_{k-1}) \quad (5)$$

$$\text{trigrams: } P(w_k|w_1^{k-1}) \approx P(w_k|w_{k-1} w_{k-2}) \quad (6)$$

$$n\text{-grams: } P(w_k|w_1^{k-1}) \approx P(w_k|w_{k-n+1}^{k-1}) \quad (7)$$

In this work, we used the SRI Language Modeling Toolkit (Stolcke 2002), a standard package commonly used by the computational linguistics and speech processing communities. Standard settings were used: Good-Turing discounting (Chen and Goodman 1996) and Katz backoff for smoothing (Katz 1987). For building the language models, we used both the filtered session data and the episode data, as summarized in Table 2. The datasets were divided in the following manner: for the filtered session dataset, a 400/76k training/test split; for the episode dataset, a 300/73k split. We varied the order of n-grams from 2-gram (bigrams) up to 8-gram.

In our first experiment, we built language models on training data and then evaluated these models on held-out test data. Cross entropy is frequently used as a metric:

$$\frac{1}{N_T}\sum_{i=1}^{n} - \log_2 P_m(t_i) \tag{8}$$

where $P_m(t_i)$ denotes the probability assigned by language model m and where the test data T is composed of sequences $(t_1, t_2, \ldots, t_n)$ containing a total of N_T symbols. The cross entropy is inversely related to the average probability a model assigns to sentences in the test data, and it is accepted that lower entropy models are preferable. Following common practices of the speech community, we report perplexity instead of cross entropy (H), where perplexity is 2^H. One advantage is that the measure has an intuitive reading: a perplexity of k means that one is as surprised on average as one would have been guessing from k equiprobable choices at each symbol in the sequence.

Results for the perplexity experiments are shown in Figure 3 for both filtered session data and episode data. We see that perplexity does decrease with higher-order models, although it appears that nothing is gained beyond 6-grams, and perplexity actually increases slightly due to the increasing sparsity of higher-order *n*-grams. Nevertheless, these results do suggest that there is regularity in sequences of user actions, and that this regularity can be modeled by taking into account history of previous actions.

In our second experiment, we applied the language models to a sequence prediction task. That is, after observing the user's current history of actions, can the model "guess" what the user is going to do next? The prediction is performed as follows: given a sequence of actions, the system generates eight sequences, one corresponding to each possible next user action. These sequences are then scored by the language model; the user action that generates the most probable sequence is then predicted.

The experimental runs were structured in the following manner: a sequence of *n*user actions (either from a single session or episode, depending on the dataset) is divided into $n - 1$ trials. At each trial, the system's task is to predict the next symbol, provided the history. Note that this experimental setup places higher-order*n*-gram models at a disadvantage, since for the first few predictions of any session or episode, information learned from higher-order *n*-grams cannot be exploited. However, this procedure does have the advantage that the number of trials is constant across all models.

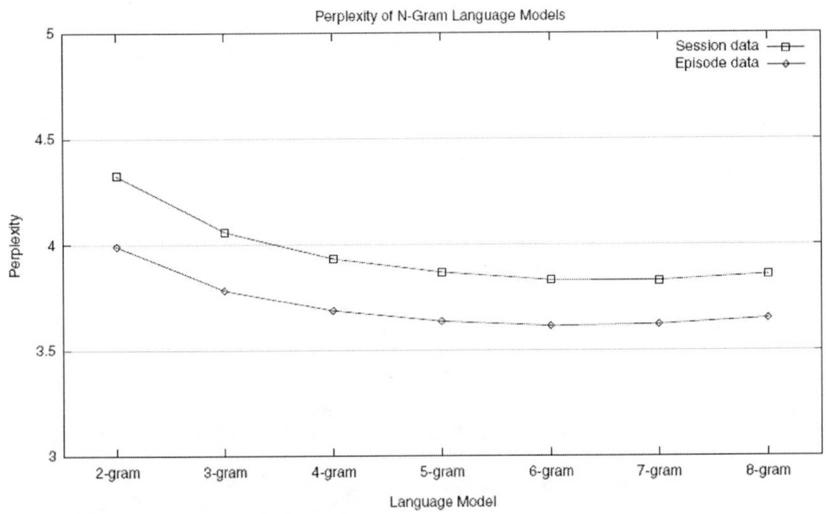

Figure 3. Perplexity of session and episode test data on different *n*-gram language models

Experiments were conducted using the language models built from the training data. From the session test data, we selected 10k sessions for testing prediction accuracy, which yielded 154,333 trials. From the episode data, we selected 20k episodes, which yielded 213,107 trials. We measured the accuracy of the predictions by comparing system output with the actual user actions—these results are shown in Figure 4. The error bars denote the 99% confidence intervals, as computed by the Clopper-Pearson method for calculating exact binomial confidence intervals (Clopper and Pearson 1934). Due to the large number of trials, the confidence intervals are quite small, thus allowing us to discriminate small differences. In both graphs, the solid line indicates the baseline (frequency of most common class: 0.388 for episode, 0.397 for sessions).

Results from the prediction accuracy experiment are consistent with the perplexity evaluation. For both session and episode data, 5-gram models significantly outperform 4-gram and lower order models. The differences in prediction accuracy between 5-, 6-, 7-, and 8-gram models are not statistically significant for both session and episode data. To facilitate comparison, the scales of the bar graphs in Figure 4 are the same. We can see that relative to the baseline, prediction accuracy is slightly higher on episode data than it is on session data: for example, 6-gram prediction accuracy for episodes is 0.503 or 30% above the baseline, while 6-gram prediction accuracy for sessions is 0.508 or 28% above the baseline. Recall that, by construction, the episode data contained more "interesting" sequences of user activities—these results suggest that there is *some* value in the episodes generated by our simple session segmentation technique.

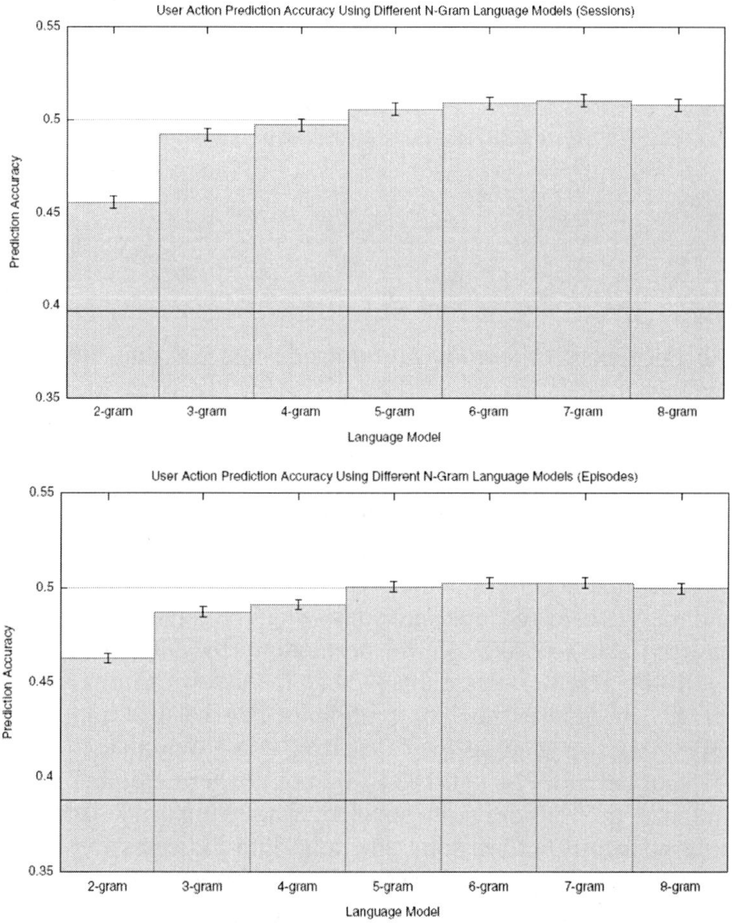

Figure 4. Accuracy of predicting next user action using different *n*-gram language models: session data on top, episode data on bottom. Solid line in each graph indicates baseline (most frequent class)

5 DETECTING ACTIVITY COLLOCATES

By encoding users' actions as string sequences, we can leverage well-known natural language processing techniques to identify interesting patterns. Computational linguists have devised a number of techniques for extracting collocations, or commonly-occurring patterns of words; see Pearce (2002) for a survey. Can such techniques be applied to help us better understand user behavior?

First, what would collocations in activity sequences derived from our logs mean? Collocations in natural language are co-occurring words that appear more frequently than one would expect by chance; often, their semantics are non-compositional (e.g., compound nouns, verb-particle constructions, etc.). In our context, collocations represent sequences of actions that are "surprising" and "meaningful". These "activity collocates" might encapsulate fragments of information-seeking strategies or "idioms" that PubMed searchers have adopted, and analyzing such sequences could reveal interesting insights about user behavior. Depending on the absolute frequency of occurrence, these information-seeking strategies might be relatively common, or exclusively used by a small population.

To find activity collocates, we used Pointwise Mutual Information (Church and Hanks 1990) to score *n*-grams from our language models:

$$\mathrm{PMI}(a_1, a_2, \ldots, a_n) = \log \frac{p(a_1, a_2, \ldots, a_n)}{p(a_1)p(a_2)\ldots p(a_n)}$$

That is, PMI quantifies the probability of seeing a particular sequence relative to the probability that the individual symbols making up the sequence co-occurred randomly. We opted for PMI instead of more complex formulations such as the log likelihood ratio (Dunning 1993) since our alphabet is small and thus our modeling task is not plagued by problems associated with rare events; cf. (Moore 2004).

Table 3 shows the top five activity collocates in terms of PMI for 2-, 3-, 4-action sequences. Analysis was performed on the 400k training samples in the filtered session data, but the episode data gives rise to similar results. The table also shows the log probability of each *n*-gram to quantify the prevalence of that particular pattern. For reference, the most frequently-occurring patterns of activity are shown in Table 4. Note that in both cases, we discarded patterns that involved non-

MEDLINE actions (the symbol 'X'), since the probabilities of those actions are distorted by our data preparation process.

Table 3. Top five 2-, 3-, and 4-sequence activity collocates, ranked by PMI value

Sequence	Count	log p	PMI
L L	100447	−1.77	1.08
L M	10778	−2.74	0.84
P P	53218	−2.05	0.80
N N	224020	−1.42	0.71
M M	1258	−3.67	0.64
L L L	55087	−2.00	2.27
L L M	5337	−3.02	1.99
M M M	162	−4.53	1.94
P P P	23421	−2.37	1.90
M R M	4225	−3.12	1.60
M M M M	44	−5.07	3.56
L L L L	33315	−2.19	3.51
L L L M	3002	−3.24	3.20
P P P P	14128	−2.57	3.14
L M M M	33	−5.20	2.71

L = related link, M = more links, N = next, P = other MEDLINE

In the PubMed interface, the detailed view of a MEDLINE citation (which contains bibliographic information, and in most cases, abstract text) is accompanied by five links to related articles, as suggested by a probabilistic content-similarity algorithm (Lin and Wilbur 2007). Below the five titles is an option to see a longer list of related articles. Action 'L' represents a click on one of these suggested article titles; action 'M' represents a click on the option to see a longer list of related articles. Overall, 'L' and 'M' represent 3.7% and 0.7% of all transactions, respectively, yet most of the activity collocates involve these two actions. That is, these sequences occur far more frequently than one would expect by chance. Consecutive L's indicate that the user was clicking on related article suggestions repeatedly (navigating from citation to citation); consecutive M's indicate that the user was navigating from lists of related articles to other lists of related articles (essentially, using MEDLINE citations repeatedly as "queries" to retrieve lists of similar citations).

Table 4. Top five most frequently-observed 2-, 3-, and 4-sequences of actions, ranked by absolute frequency

Sequence	Count	log p	PMI
R R	1108739	−0.73	0.07
Q Q	905413	−0.82	0.15
Q R	729704	−0.91	−0.03
R Q	670231	−0.95	−0.06
N N	224020	−1.42	0.71
R R R	605996	−0.96	0.24
Q Q Q	497841	−1.05	0.40
R Q R	281937	−1.29	−0.01
Q R R	271682	−1.31	−0.03
Q Q R	255293	−1.34	0.03
R R R R	379589	−1.14	0.46
Q Q Q Q	305904	−1.23	0.70
Q R R R	124797	−1.62	0.06
Q Q Q R	119846	−1.64	0.21
R Q Q Q	109770	−1.68	0.17

Q = query, R = retrieve, N = next

Our analysis suggests that browsing related articles (in its two forms) represents a*deliberate* information-seeking strategy, or more generally, a distinct way of using PubMed. Furthermore, once users begin browsing the collection in this manner, they are likely to continue—as shown by the long sequences of 'L' and 'M' actions in Table 3.

Complementary evidence presented in Lin et al. (2008) supports an even stronger claim, that the feature is indeed *useful* and *effective* for information seeking. Through analysis of document networks connected via content-similarity links, Lin et al. showed that for typical information needs, relevant documents tend to cluster together. Therefore, a user can navigate from relevant document to relevant document via these links. Finally, these results are consistent with previous studies that demonstrate the effectiveness of content-similarity browsing in simulated environments (Wilbur and Coffee 1994; Smucker and Allan 2006; Lin and Smucker 2008). Although similar features are available in many Web search engines, we are not aware of any published evidence regarding their effectiveness.

The other activity collocates are comprised of N's (next page) and P's (other MEDLINE actions). The "N N" sequence suggests that users are often persistent in examining the retrieved set—that is, they browse through at least two pages of PubMed results (each of which contains 20 citations). A natural explanation is the recall-oriented nature of typical tasks that PubMed users engage in, e.g., physicians searching for clinical evidence or biologists searching for relevant literature. This behavior stands in contrast with general Web search, where users are much more cursory in their consumption of search results. For example, Jansen et al. (2000) observed that 58% of users don't look past the first page of results (10 hits), and only around 10% of users view results past the fourth page.

As described in Table 1, the symbol 'P' serves as a catch-all category for otherwise uncategorized MEDLINE-related actions. These include actions in the "Preview/Index", "History", "Clipboard", and "Details" tabs of the PubMed interface, as well as clicks on the "LinkOut" resource supplied by external providers. Examination of the logs reveals that sequences of P's represent activities of advanced PubMed users. For example, the "Preview/Index" tab allows the user to see the number of hits that would be retrieved by a particular query—an important feature for Boolean retrieval since the result set size is often difficult to control. The "History" tab allows the user to revisit previously-issued queries. Overall, such actions are rare (only 3.8% of all page views in the filtered session dataset), but we observe consecutive P's having high PMI. As with sequences of consecutive L's and M's, we believe that this represents a distinct mode of information seeking—for example, an advanced user encountering difficulty in choosing good search terms might switch to the "Preview/Index" tab as a tool for assisting in the query formulation process.

Focusing on Table 4, it is not surprising that the most frequent patterns of activity comprise mostly of 'Q' (query) and 'R' (retrieve) actions. Together, they account for nearly three quarters of total transactions. Sequences of these two actions represent perhaps the "core" of information-seeking behavior: issuing queries, examining results, and reformulating previous queries. We also note that activity collocates (i.e., those in Table 3) aren't necessarily rare in terms of absolute frequency. For example, "N N" is the 5th most frequent 2-gram and "L L" is the 13th most frequent 2-gram.

Some frequent *n*-grams actually have surprisingly high PMI values. For example, we observe the prevalence of long query sequences. One possible interpretation is that at least some users experience difficulty formulating Boolean queries. Since almost all commercial Web search engines implement some sort of best-match algorithm, users have grown accustomed to using ranked retrieval systems. In contrast, the query formulation process in PubMed may feel quite foreign. In fact, based on analysis of a separate set of logs, we found that approximately a fifth of all PubMed queries return zero results. Related to these challenges is the difficulty associated with controlling the result set size, which is another characteristic of Boolean retrieval. For example, adding an additional term to a query that retrieves 1000 hits might yield 0 hits (this helps to explain the usefulness of the "Preview/Index" tab in the PubMed interface, for searchers who are aware of the functionality). These issues point to query formulation aids as features that could potentially benefit many PubMed users.

The sequences of consecutive repeated actions in both Tables 3 and 4 suggest an interesting behavior pattern: once a user commits to an action, he or she is likely to repeat the same action again. We can quantify this by computing $P(L|L)/P(L)$; that is, how much more likely is action 'L' to follow another action 'L' (compared to chance)? The answer in this case is around 12 times more likely: 'L' follows another 'L' about 45% of the time. We computed similar values for longer sequences and for other actions, the results of which are shown in Figure 5. Given the above discussion, the plots for action 'L' and action 'N' (next result page) are not unexpected: once users begin browsing related articles or pages in the result set, they are likely to continue doing so. However, we also observe the same effect for Query and Retrieve (albeit the effects are more minor): given a consecutive sequence of four 'Q' or 'R' actions, we are 2.07 and 1.86 more likely to see another 'Q' or 'R', respectively. This appears to support our characterization of PubMed users as "persistent", which makes sense given the complex nature of their information needs.

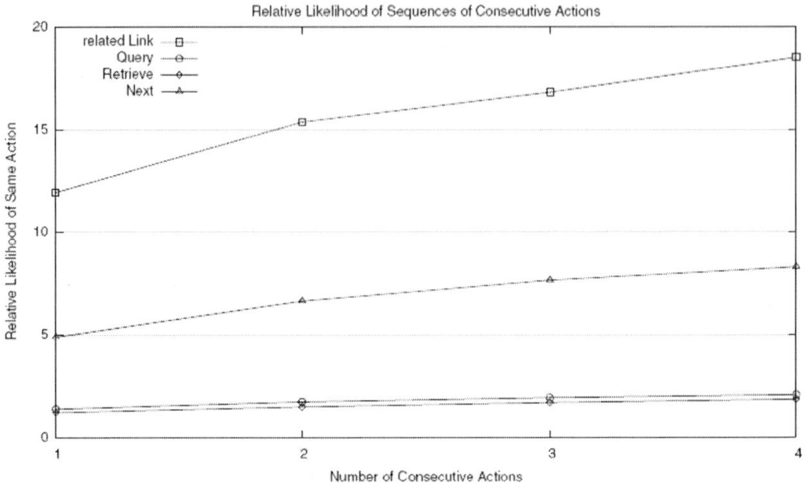

Figure 5. Relative likelihood of observing a particular action after a consecutive sequence of the same action. For example, the probability of 'L' followed by another 'L' is 12 times higher than expected by chance

6 CONCLUSION

In this work, we demonstrate that simple *n*-gram models can capture regularities in users' activities, based only on a rough characterization of their actions. With these techniques, we are able to identify activity collocates and characterize interesting patterns of behavior. Overall, the models are able to correctly predict the user's next action about half the time. We find this to be promising, considering the impoverished input—no temporal information (beyond pre-segmented sequences), no information about the semantics of the user's actions (e.g., queries they issued, the contents of the pages they clicked on, etc.). Of course, models can be enriched by exactly such data to provide a more accurate picture of users' behavior. However, the biggest barrier to this is not technological, but rather one related to policy: concerns over invasion of privacy and release of personally-sensitive information. For a discussion, see Murray and Teevan (2007). In this respect, log-mining techniques that are able to extract information from impoverished datasets are particularly valuable, since they alleviate these concerns. For academic researchers, it is more likely that such datasets can be shared.

While user modeling may be interesting in itself, our ultimate goal is to exploit such models to improve the search experience. We believe that user models can be used in three major ways:

- *Predictively.* In anticipation of what the user is likely to do next, the system can provide customized assistance. One might consider this as a generalization of query suggestion techniques that have been proposed (Anick 2003; Cui et al.2003). That is, in addition to suggesting related search terms, the system might offer helpful *actions.* For example, activity collocates might be generalized into a library of "search strategies" and presented as search aids. For example, the system may suggest something like, "similar users have found related article links to be helpful—perhaps you might want to consider..."
- *Retrospectively.* Computational models can help systems automatically cluster and classify user behavior. Results in Sect. 4 offer some possibilities for how this might be accomplished. The biggest application is user profiling and demographic modeling, which is highly relevant to many tasks ranging from personalized search (Eirinaki and Vazirgiannis2003; Shen et al. 2005) to targeted advertisement.
- *Prescriptively.* That is, results of log-based studies could be used as a basis for educating users on effective search strategies. This is not an unrealistic scenario in the context of PubMed: due to the nature of its users and their work, PubMed searchers are often willing to learn effective search techniques and advanced features.[2]

We believe that this work represents an enabler for such advances. Transaction logs help us better understand how PubMed users search for information and provide a valuable resource for building computational models of user activities. We demonstrated how user actions can be encoded as string sequences and captured with *n*-gram language models. The application of natural language processing techniques to tackle this information retrieval problem provides an example of how the two fields might productively collaborate.

FOOTNOTES

[1]http://www.nlm.nih.gov/pubs/factsheets/medline.html.

[2]Empirical evidence for this claim is demonstrated by the numerous tutorials and mini-courses offered on PubMed, as any casual Web search will reveal.

ACKNOWLEDGMENTS

We'd like to thank three anonymous reviewers for their helpful comments. This work was supported by the Intramural Research Program of the NIH, National Library of Medicine. The first author would also like to thank Esther and Kiri for their kind support.

REFERENCES

1. Agichtein, E., Brill, E., & Dumais, S. (2006). Improving Web search ranking by incorporating user behavior information. In Proceedings of the 29th Annual International ACM SIGIR Conference on Research and Development in Information Retrieval (SIGIR 2006) (pp. 19–26). Seattle, WA.

2. Anick, P. (2003). Using terminological feedback for Web search refinement—A log-based study. In Proceedings of the 26th Annual International ACM SIGIR Conference on Research and Development in Information Retrieval (SIGIR 2003) (pp. 88–95). Toronto, Canada.

3. Beitzel, S. M., Jensen, E. C., Chowdhury, A., Grossman, D., & Frieder, O. (2004). Hourly analysis of a very large topically categorized Web query log. In Proceedings of the 27th Annual International ACM SIGIR Conference on Research and Development in Information Retrieval (SIGIR 2004) (pp. 321–328). Sheffield, UK.

4. Broder, A. (2002). A taxonomy of Web search. SIGIR Forum, 36(2), 3–10.

5. Cahan, M. A. (1989). GRATEFUL MED: A tool for studying searching behavior. Medical Reference Services Quarterly, 8(4), 61–79.

6. Catledge, L. D., & Pitkow, J. E. (1995). Characterizing browsing strategies in the World-Wide Web. Computer Networks and ISDN Systems, 27(6), 1065–1073.

7. Chen, H.-M., & Cooper, M. D. (2002). Stochastic modeling of usage patterns in a Web-based information system.Journal of the American Society for Information Science and Technology, 53(7), 536–548.

8. Chen, S. F., & Goodman, J. (1996). An empirical study of smoothing techniques for language modeling. InProceedings of the 34th Annual Meeting of the Association for Computational Linguistics (ACL 1996) (pp. 310–318). Santa Cruz, CA.

9. Church, K. W., & Hanks, P. (1990). Word association norms, mutual information, and lexicography. Computational Linguistics, 16(1), 22–29.

10. Clopper, C. J., & Pearson, E. S. (1934). The use of confidence or fiducial limits illustrated in the case of the binomial. Biometrika, 26.404–413.

11. Cui, H., Wen, J.-R., Nie, J.-Y., & Ma, W.-Y. (2003). Query expansion by mining user logs. IEEE Transactions on Knowledge and Data Engineering, 15(4), 829–839.

12. De Groote, S. L., & Dorsch, J. L. (2003). Measuring use patterns of online journals and databases. Journal of the Medical Library Association, 91(2), 231–240.

13. Dunning, T. (1993). Accurate methods for the statistics of surprise and coincidence. Computational Linguistics, 19(1), 61–74.

14. Eirinaki, M., & Vazirgiannis, M. (2003). Web mining for Web personalization. ACM Transactions on Internet Technology, 3(1), 1–27.

15. Haynes, R. B., Wilczynski, N., McKibbon, K. A., Walker, C. J., & Sinclair, J. C. (1994). Developing optimal search strategies for detecting clinically sound studies in MEDLINE. Journal of the American Medical Informatics Association, 1(6), 447–458.

16. He, D., & Göker, A. (2000). Detecting session boundaries from Web user logs. In Proceedings of the BCS-IRSG 22nd Annual Colloquium on Information Retrieval Research (pp. 57–66). Cambridge, UK.

17. Hersh, W. R., Cohen, A., Ruslen, L., & Roberts, P. (2007). TREC 2007 Genomics Track overview. InProceedings of the Sixteenth Text REtrieval Conference (TREC 2007). Gaithersburg, MD.

18. Hersh, W. R., Cohen, A., Yang, J., Bhupatiraju, R., Roberts, P., & Hearst, M. (2005). TREC 2005 Genomics Track overview. In Proceedings of the Fourteenth Text REtrieval Conference (TREC 2005), Gaithersburg, MD.

19. Herskovic, J. R., Tanaka, L. Y., Hersh, W. R., & Bernstam, E. V. (2007). A day in the life of PubMed: Analysis of a typical day's query log. Journal of the American Medical Informatics Association, 14(2), 212–220.

20. Horowitz, G. L., Jackson, J. D., & Bleich, H. L. (1983). PaperChase. Self-service bibliographic retrieval. JAMA, 250(18), 2494–2499.

21. Jansen, B. J., & Spink, A. (2004). An analysis of documents viewing patterns of Web search engine users. In A. Scime (Ed.), Web mining: Applications and techniques (pp. 339–354). Hershey, PA: IGI Publishing.

22. Jansen, B. J., & Spink, A. (2006). How are we searching the World Wide Web? A comparison of nine search engine transaction logs. Information Processing and Management, 42(1), 248–263.

23. Jansen, B. J., Spink, A., & Saracevic, T. (2000). Real life, real users, and real needs: A study and analysis of user queries on the Web. Information Processing and Management, 36(2), 207–227.

24. Joachims, T., Granka, L., Pan, B., Hembrooke, H., Radlinski, F., & Gay, G. (2007). Evaluating the accuracy of implicit feedback from clicks and query reformulations in Web search. ACM Transactions on Information Systems, 25(2), 1–27.

25. Katz, S. M. (1987). Estimation of probabilities from sparse data for the language model component of a speech recognizer. IEEE Transactions on Acoustics, Speech, and Signal Processing, 35(3), 400–401.

26. King, N. S. (1991). Search characteristics and the effects of experience on end users of PaperChase. College and Research Libraries, 52(4), 360–374.

27. Lin, J., DiCuccio, M., Grigoryan, V., & Wilbur, W. J. (2008). Navigating information spaces: A case study of related article search in PubMed. Information Processing & Management, 44(5), 1771–1783.

28. Lin, J., & Smucker, M. D. (2008). How do users find things with PubMed? Towards automatic utility evaluation with user simulations. In Proceedings of the 31st Annual International ACM SIGIR Conference on Research and Development in Information Retrieval (SIGIR 2008) (pp. 19–26). Singapore.

29. Lin, J., & Wilbur, W. J. (2007). PubMed related articles: A probabilistic topic-based model for content similarity.BMC Bioinformatics, 8, 423.

30. Manning, C. D., & Schütze, H. (1999). Foundations of statistical natural language processing. Cambridge, MA: MIT Press.

31. Moore, R. C. (2004). On log-likelihood-ratios and the significance of rare events. In Proceedings of the 2002 Conference on Empirical Methods in Natural Language Processing (EMNLP 2002) (pp. 333–340). Barcelona, Spain.

32. Murray, G. C., Lin, J., & Chowdhury, A. (2006). Action modeling: Using language models to predict query behavior. In Proceedings of the 29th Annual International ACM SIGIR Conference on Research and Development in Information Retrieval (SIGIR 2006) (pp. 681–682). Seattle, WA.

33. Murray, G. C., & Teevan, J. (2007). Query log analysis: Social and technological challenges. SIGIR Forum, 41(2), 112–120.

34. Pearce, D. (2002). A comparative evaluation of collocation extraction techniques. In Proceedings of the Third International Conference on Language Resources and Evaluation (LREC 2002) (pp. 1530–1536). Canary Islands, Spain.

35. Rose, D. E., & Levinson, D. (2004). Understanding user goals in Web search. In Proceedings of the Thirteenth International World Wide Web Conference (WWW 2004) (pp. 13–19). New York, NY.

36. Shen, X., Tan, B., & ChengXiang, Z. (2005). Implicit user modeling for personalized search. In Proceedings of the 2005 ACM CIKM International Conference on Information and Knowledge Management (pp. 824–831). Bremen, Germany.

37. Silverstein, C., Marais, H., Henzinger, M., & Moricz, M. (1999). Analysis of a very large Web search engine query log. SIGIR Forum, 33(1), 6–12.

38. Smucker, M. D., & Allan, J. (2006). Find-similar: Similarity browsing as a search tool. In Proceedings of the 29th Annual International ACM SIGIR Conference on Research and Development in Information Retrieval (SIGIR 2006)(pp. 461–468). Seattle, WA.

39. Stolcke, A. (2002). SRILM—An extensible language modeling toolkit. In Proceedings of the 7th International Conference on Spoken Language Processing (ICSLP 2002) (pp. 901–904). Denver, CO.

40. Wilbur, W. J., & Coffee, L. (1994). The effectiveness of document neighboring in search enhancement. Information Processing and Management, 30(2), 253–266.

5

EVALUATING THE PERFORMANCE OF INFORMATION RETRIEVAL SYSTEMS USING TEST COLLECTIONS

Paul Clough[1], Mark Sanderson[2]

[1]Information School, Univeristy of Sheffield, Regent Court, 211 Portobello Street, Sheffield, S10 2TN

[2]School of Computer Science and Information Technology, RMIT University, GPO Box 2476, Melbourne 3001, Victoria, Australia

ABSTRACT

Introduction

Evaluation is highly important for designing, developing and maintaining effective information retrieval or search systems as it allows the measurement of how successfully an information retrieval system meets its goal of helping users fulfil their information needs. But what does it mean to be successful? It might refer to whether an information retrieval system retrieves relevant (compared with non-relevant) documents; how quickly results are returned; how well the system supports users' interactions; whether users are satisfied with the results; how easily users can use the system; whether the system helps users carry out their tasks and fulfil their information needs; whether the system impacts on the wider environment; how reliable the system is etc. Evaluation of information retrieval systems has been actively researched for over 50 years and continues to be an area of discussion and controversy.

Test collections

In this paper we discuss system-oriented evaluation that focuses on measuring system effectiveness: how well an information retrieval system can separate relevant from non-relevant documents for a given user query. We discuss the construction and use of standardised benchmarks - test collections - for evaluating information retrieval systems.

Research directions

The paper also describes current and future research directions for test collection-based evaluation, including efficient gathering of relevance assessments, the relationship between system effectiveness and user utility, and evaluation across user sessions.

Conclusions

This paper describes test collections which have been widely used in information retrieval evaluation and provide an approach for measuring system effectiveness.

1. INTRODUCTION TO INFORMATION RETRIEVAL EVALUATION

To evaluate means to "ascertain the value or amount of something or to appraise it" (Kiewitt 1979: 3) and involves identifying suitable success criteria that can be measured in some way. Evaluation is important for designing, developing and maintaining effective Information Retrieval (or search) systems as it enables the success of an information retrieval system to be quantified and measured. This can involve evaluating characteristics of the information retrieval system itself, such as its retrieval effectiveness, or assessing consumers' acceptance or satisfaction with the system (Taube 1953). How to conduct information retrieval system evaluation has been an active area of research for the past 50 years and the subject of much discussion and debate (Saracevic 1995, Robertson 2008, Harman 2011). This is due, in part, of the need to incorporate users and user interaction into evaluation studies and the relationship between results of laboratory-based vs. operational tests (Robertson and Hancock-Beaulieu 1992).

Traditionally in information retrieval there has been a strong focus on measuring system effectiveness: the ability of an information retrieval system to discriminate between documents that are relevant or not relevant for a given user query. This focus on the system has, in part, been influenced by the focus of the information retrieval community on the development of retrieval algorithms, together with the organization of large information retrieval evaluation events, such as the Text REtrieval Conference (or TREC) in the USA. Such events have also focused on measuring system effectiveness in a controlled experimental setting (Robertson 2008, Voorhees and Harman 2005). Efficiency of an information retrieval system has also been assessed, e.g. measuring how long the system takes to return results or memory/disk space required to store the index. Measuring the effectiveness and efficiency of an information retrieval system has commonly been conducted in a laboratory setting, with little involvement of end users and focused on assessing the performance of the underlying search algorithms; therefore, this is commonly referred to as system-oriented evaluation.

Robertson and Hancock-Beaulieu (1992) suggest that the scope of 'system' in information retrieval has slowly broadened to include more elements of the retrieval context, such as the user or the user's environment, which must be included in the evaluation of information retrieval systems. Similar remarks have been made by Saracevic (1995) and Ingwersen and Järvelin (2005). Therefore, instead of focusing on just the system (i.e., its inputs and outputs), a more user-oriented approach can be taken. This may take into account the user, the user's context and situation, and their interactions with an information retrieval system, perhaps in a real-life operational environment (Borlund 2009). This includes, for example, assessing the usability of the search interface or measuring aspects of the user's information searching behaviour (e.g., a user's satisfaction with the search results or the number of items viewed/saved). Taking into account the wider context of search is part of interactive information retrieval evaluation (Kelly 2009).

In practice it is common to utilize various evaluation approaches that will be used throughout the development of an information retrieval system, from using test collections to develop, contrast and optimize search algorithms; to conducting laboratory-based user experiments for improving design of the user interface; to evaluation carried out in situ as the information retrieval system is used in a practical setting. Järvelin (2011) contrasts system- and user-oriented approaches to information retrieval evaluation in more controlled and artificial settings with the evaluation of information retrieval systems in operational real-life settings, where it is possible to employ both user- and system-oriented approaches. The rest of this chapter focuses on system-oriented approaches to information retrieval evaluation using test collections, also referred to as the Cranfield approach.

2. EVALUATION OF INFORMATION RETRIEVAL SYSTEMS USING TEST COLLECTIONS

For decades the primary approach to evaluation has been system-oriented, focusing on assessing how well a system can find documents of interest given a specification of the user's information need. One of the most used methodologies for conducting experiments that can be repeated and conducted in a controlled laboratory-based setting is test collection-based evaluation (Robertson 2008, Sanderson 2010, Järvelin 2011, Harman 2011). This approach to evaluation has its origins in experiments conducted at Cranfield library in the UK, which ran between 1958 and 1966, and is often referred to as the Cranfield approach or methodology (Cleverdon 1991). The aim of the Cranfield experiments was to create "a laboratory type situation where, freed as far as possible from the combination of operational variables, the performance of index languages could be considered in isolation" (Cleverdon 1967). This approach defined the need for a common document collection, query tasks and ground truths to evaluate different indexing strategies under controlled conditions abstracted away from the operational environment.

The Cranfield approach to information retrieval evaluation uses test collections: re-useable and standardised resources that can be used to evaluate information retrieval systems with respect to system. The main components of an information retrieval test collection are the document collection, topics, and relevance assessments. These, together with evaluation measures, simulate the users of a search system in an operational setting and enable the effectiveness of an information retrieval system to be quantified. Evaluating information retrieval systems in this manner enables the comparison of different search algorithms and the effects on altering algorithm parameters to be systematically observed and quantified. Although proposed in the 1960s, this approach was popularised through the NIST-funded TREC series of large-scale evaluation campaigns that began in 1992, and subsequently other large-scale evaluation campaigns (e.g., CLEF, NTCIR and FIRE), that have stimulated significant developments in information retrieval over the past 20 years (Voorhees and Harman 2005).

The most common way of using the Cranfield approach is to compare various retrieval strategies or systems, which is referred to as comparative evaluation. In this case the focus is on the relative performance between systems, rather than absolute scores of system effectiveness. To evaluate using the Cranfield approach typically requires these stages: (1) select different retrieval strategies or systems to compare; (2) use these to produce ranked lists of documents (often called *runs*) for each query (often called topics); (3) compute the effectiveness of each strategy for every query in the test collection as a function of relevant documents retrieved; (4) average the scores over all queries to compute overall effectiveness of the strategy or system; (5) use the scores to rank the strategies/systems relative to each other. In addition, statistical tests may be used to determine whether the differences between effectiveness scores for strategies/systems and their rankings are significant. This is necessary if one wants to determine the 'best' approach. In the TREC-style version of the Cranfield approach, there is a further stage required prior to (2) above, whereby the runs for each query are used to create a pool of documents that are judged for relevance, often by domain experts. This produces a list of relevant documents (often called *qrels*) for each query that is required in computing system effectiveness with relevance-based measures (e.g., precision and recall).

Test collection-based evaluation is highly popular as a method for developing retrieval strategies. Benchmarks can be used by multiple researchers to evaluate in a standardised manner and with the same experimental set up, thereby enabling the comparison of results. In addition, user-oriented evaluation, although highly beneficial, is costly and complex and often difficult to replicate. It is this stability and standardization that makes the test collection so attractive. However, there are a number of limitations to test collection-based evaluation due to its abstraction from reality (Ingwersen and Järvelin 2005, pp. 6-9). Test collections experiments make a number of assumptions: that the relevance of documents is independent of each other; that all documents are equally important; that the user's information need remains static; that a single set of judgments for a query is representative of the user population; that the lists of relevant documents for each query are exhaustive (Voorhees 2002).

2.1 Building test collections

A test collection usually consists of a document collection, a set of topics that describe a user's information need and a set of relevance judgments indicating which documents in the collection are relevant to each topic. When constructing a test collection there are typically a number of practical issues that must be addressed (Sanderson and Braschler 2009). By modifying the components of a test collection and evaluation measures used, different retrieval problems and domains can be simulated. The original and most common problem modelled is ad hoc retrieval: the situation in which an information retrieval system is presented with a previously unseen query. However, test collection-based evaluations have also been carried out on tasks including question answering, information filtering, text summarization, topic detection and tracking, image and video retrieval, and text summarization.

2.2 Document collection

IR systems index documents that are retrieved in response to users' queries. A test collection must contain a static set of documents that should reflect the kinds of documents likely to be found in the operational setting or domain. This might involve digital library collections or sets of Web pages; texts or multimedia items (e.g., images and videos). The notion of a static document collection is important as it ensures that results can be reproduced upon re-use of the test collection. Specific questions that might be considered when gathering documents include:

- **How many items should be gathered?** In the case of smaller digital library collections, all items could be included in the test collection. However, in situations involving larger collections or constantly changing collections, e.g. the Web, then a sample of documents must be collected.

- **What items should be sampled to create the document collection?** Items in the test collection should represent as faithfully as possible those found in an operational setting. If items must be selected for a document collection then it may be appropriate to select a random sample or a specific subset of documents.
- **What about copyright constraints?** If the document collection is to be shared with others then an important aspect of which documents to include may involve considering copyright issues.

2.3 Topics

Information retrieval systems are evaluated for how well they answer users' search requests. In the case of ad hoc retrieval, the test collection must contain a set of statements that describe typical users' information needs. These might be expressed as queries that are submitted to an information retrieval system, questions or longer written descriptions. For example, TREC uses the notion of a *topic*, which typically consists of three fields: query, title and description. The query field represents a typical set of keywords a user might issue for a given topic. The title field provides a longer description, normally a sentence, of an information need. The description field describes in more detail a user's information and what they are attempting to find. This often includes a description of what constitutes relevant (and non-relevant) documents and may be used by people judging relevance (if not the person who generated the topic). Practical issues that often arise during creation of topics include:

- **How should a suitable set of topics be generated?** Ideally realistic queries should be developed for the test collection. These could be derived, for example, from analysing the query logs of operational systems, from the results of conducting user studies with end users of existing systems, or based on the input of domain experts.
- **How many topics are required for obtaining reliable evaluation results?** In practice, a minimum of 50 topics should be included in

the test collection to ensure reliable. However, results have also shown that making fewer relevance judgments over greater numbers of queries leads to more reliable evaluation (Carterette *et al.* 2008).

- **Do the topics represent a diverse enough set of information needs?** Often queries on a range of topics/themes should be selected with varying characteristics to test information retrieval systems under a range of settings. For example, one might include shorter and longer queries, queries for specific entities (e.g., people or places) versus more subject-based queries. Research has shown that some queries are more useful in evaluation than others (Mizzaro and Robertson 2007).

- **How should the topics be expressed?** The topics might be expressed as a set of keywords and/or verbal descriptions. For other forms of media this may include non-textual descriptions. For example, for image retrieval the topics might include example images that can be used to initiate retrieval. It should be noted, however, that realistic queries in practice tend to be short and ambiguous.

2.4 Relevance assessments

For each topic in the test collection, a set of relevance judgments must be created indicating which documents in the collection are relevant to each topic. The notion of relevance used in the Cranfield approach is commonly interpreted as topical relevance: whether a document contains information on the same topic as the query. In addition, relevance is assumed to be consistent across assessors and static across judgments. However, this is a narrow view of relevance, which has been shown to be subjective, situational and multi-dimensional (Schamber 1994). Some have speculated that this variability would assess the accuracy of the measurement of retrieval effectiveness. A series of experiments were conducted to test this hypothesis (Cleverdon 1970, Voorhees 1998) with results showing that despite there being marked differences in the documents that different assessors judged as relevant or non-relevant, the differences did not substantially affect the relative ordering of information retrieval systems being measured using the different assessments.

Relevance judgments can be binary (relevant or not relevant) or use graded relevance judgments, e.g. highly relevant, partially relevant or non-relevant. The use of binary versus graded relevance judgments is important as it has implications for which evaluation measures can be used to evaluate the information retrieval systems. Commonly the assessors will be the people who originally created the topics, but this is not always the case. For example, relevance assessments might be gathered in a distributed way using multiple assessors and crowdsourcing. It must, however, be recognised that domain expertise can affect the quality of relevance assessments obtained (Bailey *et al.* 2008, Kinney *et al.* 2008).

There are various ways of gathering the relevance assessments. For example, in TREC the common approach is to gather the top *n* results from the different information retrieval systems under test for each topic and aggregate results into a single list of results for judging (called *pooling*). This assumes that the result lists of different information retrieval systems are diverse and therefore will bring relevant documents into the pool. The relevance assessors then go through the pool and make relevance judgments on each document which can then be used to compute system effectiveness. Documents which are not judged are often categorised as not relevant. An issue with pooling is the completeness of relevance assessments. Ideally for each topic one should find all relevant documents in the document collection; however, pooling may only find a subset. Approaches to help overcome this include using results lists from searches conducted manually in the pool of documents for assessment, or supplementing the sets of relevance judgments with additional relevant documents discovered during further manual. Generating complete sets of relevance judgments helps to ensure that when evaluating future systems, improvements in results can be detected. Generating relevance assessment is often highly time-consuming and labor intensive. This often leads to a bottleneck in the creation of test collections. Various techniques have been proposed to make the process of relevance assessment more efficient. Practical issues that often arise during relevance assessments include:

- **Who should do the assessments?** Ideally the topic generator will also perform relevance judgments. Bailey *et al.* (2008) compared the relevance assessments of judges who were subject experts and those who were not. They found that different assessments resulted, which had an effect on the measurement of system effectiveness (although the ranking remained the same).

- **How many assessments should be made?** There is evidence to suggest that assessing fewer documents for more topics (shallow and wide) is better than making exhaustive judgments for fewer topics (narrow and deep) (Carterette *et al.* 2008).
- **What are the assessors expected to do?** The relevance schemes must be decided and assessors provided with clear instructions on how to make relevance assessments. Kinney *et al.* (2008) showed that relevance assessments were more reliably obtained if detailed descriptions of what the user, issuing a query, was seeking.
- **What about finding missing relevant documents?** If a new system or configuration finds relevant documents that have not been identified previously its performance will not be maximal (especially for ad hoc retrieval and using pooling techniques). Either the original set of judgments should be as complete as possible using techniques to supplement the pools, testers should use measures of system effectiveness that can deal with incomplete judgments (Aslam *et al.* 2006), or testers should generate new judgments for results obtained from new systems.

2.5 Assessing system effectiveness: evaluation measures

Evaluation measures provide a way of quantifying retrieval effectiveness (Manning *et al.* 2008, Croft *et al.*2009). Together, the test collection and evaluation measure provide a simulation of the user of an information retrieval system. For example, in the case of ad hoc retrieval the user is modelled as submitting a single query and being presented with a ranked list of results. One assumes that the user then starts at the top of the ranked list and works their way down examining each document in turn for relevance. This, of course, is an estimation of how users behave; in practice they are often far less predictable. There are also further complications that must be considered. For example, research has shown that users are more likely to select documents higher up in the ranking (*rank bias*); measures typically assume that no connection exists between retrieved documents (*independence assumption*); and particularly in the case of Web search, one must decide how to deal with duplicate documents: should they be counted as relevant or ignored as they have been previously seen?

2.6 Set-based measures

Two simple measures developed early on were *precision* and *recall*. These are *set-based measures*: documents in the ranking are treated as unique and the ordering of results is ignored. Precision measures the fraction of retrieved documents that are relevant; recall measures the fraction of relevant documents that are retrieved. Precision and recall hold an approximate inverse relationship: higher precision is often coupled with lower recall. However, this is not always the case as it has been shown that precision is affected by the retrieval of non-relevant documents; recall is not. Compared to other evaluation measures, precision is simple to compute because one only considers the set of retrieved documents (as long as relevance can be judged). However, to compute recall requires comparing the set of retrieved documents with the entire collection, which is impossible in many cases (e.g., for Web search). In this situation techniques, such as pooling, are used.

Often preference is given to either precision or recall. For example, in Web search the focus is typically on obtaining high precision by finding as many relevant documents in the top n results. However, there are certain domains, such as patent search, where the focus is on finding all relevant documents through an exhaustive search. Alternative recall-oriented measures can then be employed (Magdy and Jones 2010). Scores for precision and recall are often combined into a single measure to allow the comparison of information retrieval systems. Example measures include the *e* and *f* measures (van Rijsbergen 1979).

2.7 Rank-based measures

More commonly used measures are based on evaluating *ranked retrieval* results, where importance is placed, not only on obtaining the maximum number of relevant documents, but also for returning relevant documents higher in the ranked list. A common way to evaluate ranked outputs is to compute precision at various levels of recall (e.g., 0.0, 0.1, 0.2, ... 1.0), or at the rank positions of all the relevant documents and the scores averaged (referred to as *average precision*). This can be computed across multiple queries by taking the arithmetic mean of average precision values for individual topics. This single-figure measure of precision across relevant documents and multiple queries is referred to as *mean average precision* (or MAP). A further common measure is precision at a fixed rank position, for example *Precision at rank 10* (P10 or P@10). Because the number of relevant documents can influence the P@10 score, an alternative measure called *R-precision* can be used: precision is measured at the rank position *Rq*, the total number of relevant documents for query *q*.

More recently, measures based on non-binary (or *graded*) relevance judgments have been utilised, such as*discounted cumulative gain* (Järvelin and Kekäläinen 2002). In such measures, each document is given a score indicating relevance (e.g., relevant=2; partially-relevant=1; non-relevant=0). Discounted cumulative gain computes a value for the number of relevant documents retrieved that includes a discount function to progressively reduce the importance of relevant documents found further down the ranked results list. This simulates the assumption that users prefer relevant documents higher in the ranked list. The measure also makes the assumption that highly relevant documents are more useful than partially relevant documents, which in turn are more useful than non-relevant documents. The score can be normalised to provide a value in the range 0 to 1, known as *normalised DCG* (nDCG). The measure can be averaged across multiple topics similar to computing mean average precision, and it has also been extended to compute the value of retrieved results across multiple queries in a session, referred to as *normalised session discounted cumulative gain* or nsDCG (Järvelin *et al.* 2008).

2.8 Other measures

Additional measures have been developed to evaluate different information retrieval problems. For example, to measure the success of search tasks where just one relevant document is required (known-item search), measures, such as *mean reciprocal rank* (MRR), can be used. Another problem requiring bespoke evaluation measures is the assessment of the variability or diversity of results. In these cases one might want to evaluate the different aspects (or sub-topics) of query and therefore measures such as *S-recall* (Zhai *et al.* 2003) were created. This measure promotes systems that return different relevant documents rather than all from the same (or similar) topic or aspect.

In practice it is important to select an evaluation measure that is suitable for the given task; for example, if the problem is known-item search then the mean reciprocal rank would be appropriate; for an *ad hoc* search task then mean average precision or averaged normalised discounted cumulative gain would be applicable.

3. COMPARING RESULTS

It is common to find in practice that retrieved results are compared against each other (comparative evaluation) with the runs with the highest scores deemed as the 'best'. In this case an absolute score of retrieval performance is of less importance than scores relative to each other. One might compare sets of results produced by running the same system multiple times with varying parameter settings, or taking single runs from multiple systems and comparing them to determine the *best* search system. Typically evaluation measures are computed across multiple queries (or topics) and averaged to produce a final score. When comparing systems, significance testing should be used to determine whether one ranking is actually better than another rather than due to chance. Commonly used tests are Kendall's Tau and the Wilcoxon signed-rank test (Sanderson 2010: 63-73).

4. CURRENT AND FUTURE RESEARCH DIRECTIONS

Laboratory-based information retrieval evaluation using test collections is a popular, and useful, approach to evaluation. However, information retrieval researchers have recognised the need to update the original Cranfield approach to allow the evaluation of new information retrieval systems that deal with varying search problems and domains (Kamps *et al.* 2009). Research is ongoing to tackle a range of issues in information retrieval evaluation using test collections and we discuss three examples in which we have been personally involved: gathering relevance assessments efficiently, comparing system effectiveness and user utility; and evaluating information retrieval systems over sessions. Further avenues of research include the use of simulations (Azzopardi *et al.* 2010), and the development of new information retrieval evaluation measures (Yilmaz *et al.*2010, Smucker and Clarke 2012).

4.1 Gathering relevance assessments efficiently

One area receiving recent attention by the information retrieval research community is how to create test collections efficiently. Given the bottleneck in gathering relevance judgments, 'low-cost evaluation' techniques have been proposed. These include approaches based on focusing assessor effort on runs from particular systems or topics that are likely to contain more relevant documents (Zobel 1998), sampling documents from the pool (Aslam *et al.* 2006), supplementing pools with relevant documents found by manually searching the document collection with an information retrieval system, known as *interactive search and judge* or ISJ (Cormack *et al.* 1998) and simulating queries and relevance assessments based on user's queries and clicks in search logs (Zhang and Kamps 2010).

One particular approach that has received interest recently has been the use of crowdsourcing: the act of taking a job traditionally performed by a designated person and outsourcing to an undefined, generally large, group of people in the form of an open call. Amazon Mechanical Turk (AMT) is one such example of a crowdsourcing platform. This system has around 200,000 workers from many countries that perform human intelligence tasks. Recent research has demonstrated that crowdsourcing is feasible for gathering relevance assessments (Alonso and Mizzaro 2009, Kazai 2011, Carvalho *et al.* 2011). However, previous studies have also demonstrated that domain expertise can have an impact on the quality and reliability of relevance judgments (Bailey *et al.* 2008, Kinney *et al.* 2008) and is particularly problematic when using crowdsourcing in specialised domains (Clough *et al.* 2012).

4.2 System effectiveness versus user utility

The effectiveness of information retrieval systems is typically measured based on the number of relevant items found. The test collection and measure aim to simulate users in an operational setting with the assumption that if System A were to score higher than System B on a test collection then users would prefer or be more satisfied with System A over System B. This is important because if this were not the case then test collection-based evaluation would be limited: improvements in system effectiveness in the laboratory would not necessarily translate to improvements in an operational setting and thereby benefit the user.

Several studies have been undertaken to investigate the link between system effectiveness and user preference or satisfaction. Results from these studies have differed widely: some have shown that an increase in system effectiveness did not have detectable gains for the end user in practice (Hersh *et al.* 2000); others have demonstrated that when comparing two systems with large relative differences in system effectiveness and user preferences for which system output they prefer, results were strongly correlated (Al-Maskari *et al.* 2008,Sanderson *et al.* 2010). Ingwersen and Järvelin (2005) assert that the real issue in information retrieval system design is not whether precision and recall goes up, but rather whether the system helps users perform search tasks more effectively. There are still many unanswered questions about the link between system- and user-oriented measures of search effectiveness and satisfaction.

4.3 Evaluation across user sessions

Test collection-based evaluation to date has focused on evaluating information retrieval systems in a single query-response manner. However, in practice users typically reformulate their queries in response to results from the information retrieval system or as their information need alters during a search episode (Bates 1989;Jansen *et al.* 2009). This requires re-thinking information retrieval evaluation to compute system success over multiple query-response interactions (Järvelin 2009, Keskustalo *et al.* 2009, Kanoulas *et al.* 2011). For example, traditional evaluation measures, such as precision and recall, are not valid for multiple queries; therefore proposals have been made to extend existing measures. For example, Järvelin *et al.* (2008) extend the*normalised discounted cumulative gain* (nDCG) measure to a measure that considered multi-query sessions called *normalised session discounted cumulative gain* (nsDCG); Kanoulas *et al.* (2011) generalize traditional evaluation measures such as average precision to multi-query session evaluation. Further aspects that must be considered are whether or not to include duplicate documents when evaluating multiple query-responses.

An example of a large-scale evaluation effort to evaluate information retrieval systems over multi-query sessions is the TREC Session Track, which has run since 2010. The TREC organisers have provided the resources and evaluation measures to assess whether systems with prior knowledge of a user's search behaviour can provide more effective search results for subsequent queries (Kanoulas *et al.* 2012). The test collection for session evaluation consists of a document collection, topics, and relevance assessments. However, the test collection for sessions also contains information from real user sessions, such as retrieved results, the items clicked on by users and the length of time spent viewing results and selected items. Results have broadly shown that information retrieval results can be improved for a given query with users' interaction data and that the more data used the higher the performance obtained.

5. SUMMARY

Like conducting any information retrieval experiment, the use of a test collection-based approach must be planned and decisions made with respect to the experimental design (Tague-Sutcliffe 1996). The goals of the evaluation must be defined; a suitable test collection must be selected from those already in existence, or must be created specifically for the retrieval problem being addressed; different information retrieval systems or techniques must be developed or chosen for testing and comparing; evaluation measures and statistical tests must be selected for evaluating the performance of the information retrieval system of for comparing whether one version of the system is better than another. These decisions are important, as they will affect the quality of the benchmark and impact on the accuracy and usefulness of results.

Over the years, test collection-based evaluation has been highly influential in shaping the core components of modern information retrieval systems, including Web search. This has been particularly visible in the context of TREC and related studies, which have not only provided the necessary benchmarks to compare different approaches to retrieval, but also provided a community and forum in which it can be discussed and promoted. The Cranfield approach to information retrieval evaluation has remained popular for more than fifty years although it is important to remember the original purpose of test collection experimentation for information retrieval: to develop and optimize algorithms for locating and ranking a set of documents about the same topic to a given query. Indexing, matching and ranking documents with respect to queries are the basic requirements of an information retrieval algorithm, which can be tested in a laboratory using test collections. There are widely recognised weaknesses to the Cranfield approach that will provide the research community with many avenues for further research in coming years.

ABOUT THE AUTHORS

Paul Clough is a Senior Lecturer in the Information School at the University of Sheffield. He received a B.Eng. (hons) degree in Computer Science from the University of York in 1998 and a Ph.D. from the University of Sheffield in 2002. His research interests mainly revolve around developing technologies to assist people with accessing and managing information. Paul can be contacted at p.d.clough@ sheffield.ac.uk

Mark Sanderson is a Professor in the School of Computer Science and Information Technology at RMIT University in Melbourne, Australia. He received his Bachelor's and Doctorate degrees in Computer Science at the University of Glasgow, Scotland. He can be contacted at mark.sanderson@rmit.edu.au

REFERENCES

1. Al-Maskari, A., Sanderson, M., Clough, P. & Airio, E. (2008). The good and the bad system: does the test collection predict users' effectiveness? In *Proceedings of the 31st Annual International ACM SIGIR Conference on Research and Development in Information Retrieval.* (pp. 59-66). New York, NY: ACM Press. Retrieved 5 May, 2013 from http://ir.shef.ac.uk/cloughie/papers/fp440-almaskari. pdf Archived by WebCite® at http://www.webcitation.org/ 6H0NCnbHd)

2. Alonso, O. & Mizzaro, S. (2009). Can we get rid of TREC assessors? Using Mechanical Turk for relevance assessment. In Shlomo Geva, Jaap Kamps, Carol Peters, Tetsuya Sakai, Andrew Trotman & Ellen Voorhees, (Eds.). *Proceedings of the SIGIR 2009 Workshop on the Future of information retrieval Evaluation.* (pp. 15-16). Amsterdam: IR Publications. Retrieved 5 May, 2013 from http:// staff.science.uva.nl/%7Ekamps/ireval/papers/paper_22.pdf (Archived by WebCite® at http://www.webcitation.org/ 6H1mV 6taR)

3. Aslam, J.A., Pavlu, V. & Yilmaz, E. (2006). A statistical method for system evaluation using incomplete judgments. In *Proceedings of the 29th Annual International ACM SIGIR Conference on Research and Development in Information Retrieval,* (pp. 541-548). New York, NY: ACM Press

4. Azzopardi, L., Järvelin, K., Kamps, J., & Smucker, M.D. (2010). . *SIGIR Forum,* **44**(2), 35-47. Retrieved 5 May, 2013 from http://www.sigir.org/forum/2010D/sigirwksp/2010d_sigirforum_ azzopardi.pdf (Archived by WebCite® at http://www.webcitation. org/6H1miejTW)

5. Bailey, P., Craswell, N., Soboroff, I., Thomas, P., de Vries, A.P. & Yilmaz, E. (2008). Relevance assessment: are judges exchangeable and does it matter? In Sung-Hyon Myaeng, Douglas W. Oard, Fabrizio Sebastiani, Tat-Seng Chua, Mun-Kew Leong (Eds.). *Proceedings of the 31st Annual International ACM SIGIR Conference on Research and Development in Information Retrieval, Singapore, July 20-24, 2008..* (pp. 667-674). New York, NY: ACM Press. Retrieved 5 May, 2013 from http://es.csiro.au/pubs/bailey_ sigir08.pdf (Archived by WebCite® at http://www.webcitation.org/ 6GOBCHiqK)

6. Bates, M.J. (1989). The design of browsing and berrypicking techniques for the online search interface. *Online Review,* **13**(5), 407-424. Retrieved 5 May, 2013 from http://pages.gseis.ucla.edu/

faculty/bates/berrypicking.html Archived by WebCite® at http://www.webcitation.org/6H1mrhwxF)

7. Borlund: (2009). User-centred evaluation of information retrieval systems. In A. Göker and J. Davies (eds),*Information retrieval: searching in the 21st Century*, Chicester, UK: John Wiley & Sons

8. Carterette, B., Pavlu, V., Kanoulas, E., Aslam, J.A. & Allan, J. (2008). Evaluation over thousands of queries. In Sung-Hyon Myaeng, Douglas W. Oard, Fabrizio Sebastiani, Tat-Seng Chua, Mun-Kew Leong (Eds.). *Proceedings of the 31st Annual International ACM SIGIR Conference on Research and Development in Information Retrieval, Singapore, July 20-24, 2008.*. (pp. 651-658). New York, NY: ACM Press. Retrieved 5 May, 2013 from http://ir.cis.udel.edu/~ carteret/papers/sigir08.pdf Archived by WebCite® at http:// www.webcitation.org/6H1n04iqc)

9. Carvalho, V. R., Lease, M. & Yilmaz, E. (2011). Crowdsourcing for search evaluation. *SIGIR Forum*, **44**(2), 17-22. Retrieved 5 May, 2013 from http://www.sigir.org/forum/2010D/sigirwksp/2010d_ sigirforum_carvalho.pdf Archived by WebCite® at http://www. webcitation.org/6H1n8yVN1)

10. Cleverdon, C.W. (1967). The Cranfield tests on index language devices. *Aslib Proceedings*, **19**(6), 173-192

11. Cleverdon, C. W. (1970). *The effect of variations in relevance assessments in comparative experimental tests of index languages.* Cranfield, UK: Cranfield Institute of Technology. (Cranfield Library Report, No. 3)

12. Cleverdon, C.W. (1991). The significance of the Cranfield tests on index languages. In Abraham Bookstein, Yves Chiaramella, Gerard Salton, Vijay V. Raghavan (Eds.). *Proceedings of 14th Annual International ACM SIGIR Conference on Research and Development in Information Retrieval, Chicago, IL, USA October 13 - 16, 1991.* (pp. 3-12). New York, NY: ACM Press

13. Clough, P., Sanderson, M., Tang, J., Gollins, T. & Warner, A. (2012). Examining the limits of crowdsourcing for relevance assessment. *IEEE Internet Computing*, In Print. Retrieved 5 May, 2013 from http://www.seg.rmit.edu.au/mark/publications/my_papers/IEEE-IC-2012.pdf Archived by WebCite® at http://www.webcitation.org/6H1nE0LNX)

14. Cormack, G.V., Palmer, C.R. & Clarke, C.L.A. (1998). Efficient construction of large test collections. In*Proceedings of the 21st Annual International ACM SIGIR Conference on Research and Development in Information Retrieval, August 24-28 1998, Melbourne, Australia.*. (pp. 282-289). New York, NY: ACM Press.

15. Croft, B., Metzler, D. & Strohman, T. (2009). *Search engines: information retrieval in practice*. Boston, MA, Pearson Education

16. Harman, D. (2011). *Information retrieval evaluation*. San Raphael, CA: Morgan & Claypool Publishers. (Synthesis Lectures on Information Concepts, Retrieval, and Services, Volume 3, No. 2).

17. Hersh, W., Turpin, A., Price, S., Chan, B., Kramer, D., Sacherek, L. & Olson, D. (2000). Do batch and user evaluations give the same results? In Nicholas J. Belkin, Peter Ingwersen, Mun-Kew Leong (Eds.). *Proceedings of the 23rd Annual International ACM SIGIR Conference on Research and Development in Information Retrieval, July 24-28, 2000, Athens, Greece.*. (pp. 17-24). New York, NY: ACM Press. Retrieved 5 May, 2013 from http://skynet.ohsu.edu/~hersh/sigir-00-batcheval.pdf Archived by WebCite® at http:// www.webcitation.org/6H1nNaYid)

18. Ingwersen: & Järvelin, K. (2005). *The turn: integration of information seeking and retrieval in context*, New York, NY: Springer-Verlag

19. Jansen, B.J., Booth, D.L., & Spink, A. (2009). Patterns of query reformulation during web searching. *Journal of the American Society for Information Science and Technology*, **60**(7), 1358-1371. Retrieved 5 May, 2013 from http://faculty.ist.psu.edu/jjansen/

academic/pubs/jansen_patterns_query_reformulation.pdf Archived by WebCite® at http://www.webcitation.org/6H1nU1bRD)

20. Järvelin, K. & Kekäläinen, J. (2002). Cumulated gain-based evaluation of information retrieval techniques. *ACM Transactions on Information Systems*, **20**(4), 422-446

21. Järvelin, K., Price, S.L., Delcambre, L.M.L. & Nielsen, M.L. (2008). Discounted cumulated gain based evaluation of multiple-query IR sessions. In *Proceedings of the 30th European Conference on Advances in Information Retrieval.* (pp. 4-15). Berlin: Springer. (Lecture Notes in Computer Science, 4956).

22. Järvelin, K. (2009). Explaining user performance in information retrieval: challenges to IR evaluation. In*Proceedings of the 2nd International Conference on Theory of Information Retrieval: Advances in Information Retrieval Theory*. (pp. 289-296). Berlin: Springer.

23. Järvelin, K. (2011). Evaluation. In Ruthven, I. and Kelly, D. (eds.), *Interactive information seeking, behaviour and retrieval.* (pp. 113-138). London, UK: Facet Publishing.

24. Kamps, J., Geva, S., Peters, C., Sakai, T., Trotman, A. & Voorhees, E. (2009). Report on the SIGIR 2009 workshop on the future of information retrieval evaluation. *SIGIR Forum*, **43**(2), 13-23. Retrieved 5 May, 2013 from http://www.sigir.org/forum/ 2009D/ sigirwksp/2009d_sigirforum_kamps.pdf Archived by WebCite® at http://www.webcitation.org/6H1nmVq7g)

25. Kanoulas, E., Carterette, B., Clough, P. & Sanderson, M. (2011). Evaluating multi-query sessions. In Wei-Ying Ma, Jian-Yun Nie, Ricardo A. Baeza-Yates, Tat-Seng Chua, W. Bruce Croft (Eds.). *Proceedings of the 34th International ACM SIGIR Conference on Research and Development in Information Retrieval*, Beijing, China July 24 - 28, 2011. (pp. 1053-1062). New York, NY: ACM Press. Retrieved 5 May, 2013 from http://ir.cis.udel.edu/~ carteret/

papers/sigir11b.pdf Archived by WebCite® at http://www.webcitation.org/6H1nvBDqm)

26. Kanoulas, E., Carterette, B., Hall, M., Clough: & Sanderson, M. (2012). Session Track 2012 Overview. In*Proceedings of the 21st Text REtrieval Conference.* Washington, DC: National Institute of Standards and Technology

27. Kazai, G. (2011). In search of quality in crowdsourcing for search engine evaluation. In *Proceedings of the 33rd European Conference on Advances in Information Retrieval.* (pp. 165-176). Berlin: Springer,

28. Kelly D. (2009). Methods for evaluating interactive information retrieval systems with users. *Foundations and Trends in Information Retrieval,* **3**(1-2), 1-224. Retrieved 5 May, 2013 from http://www.ils.unc.edu/~dianek/FnTIR-Press-Kelly.pdf Archived by WebCite® at http://www.webcitation.org/6H1o2TKSU)

29. Keskustalo, H., Järvelin, K., Pirkola, A., Sharma, T., and Lykke, M. (2009). Test Collection-Based information retrieval Evaluation Needs Extension toward Sessions - A Case of Extremely Short Queries. In *Proceedings of the 5th Asia Information Retrieval Symposium on Information Retrieval Technology,* (pp. 63-74). Berlin:Springer. Retrieved 5 May, 2013 from http://www.sis.uta.fi/infim/julkaisut/fire/2009/HK-SessEval-AIRS09-prep.pdf Archived by WebCite® at http://www.webcitation.org/6H1o7fRB8)

30. Kiewitt, E.L. (1979). *Evaluating information retrieval systems: the PROBE program.* Westport, CT: Greenwood Press

31. Kinney, K. A., Huffman, S.B. & Zhai, J. (2008). How evaluator domain expertise affects search result relevance judgments. In *Proceedings of the 17th ACM Conference on Information and Knowledge Management.* (pp. 591-598). New York, NY: ACM Press.

32. Magdy, W., & Jones, G. (2010). PRES: a score metric for evaluating recall-oriented information retrieval applications. In *Proceedings of the 33rd International ACM SIGIR Conference on Research and*

Ddevelopment in Information Retrieval, Geneva, Switzerland July 19 - 23, 2010. (pp. 611-618). New York, NY: ACM Press.

33. Manning, C.D., Raghavan: & Schütze, H. (2008). Introduction to information retrieval. Cambridge: Cambridge University Press. Retrieved 5 May, 2013 from http://nlp.stanford.edu/IR-book/ Archived by WebCite® at http://www.webcitation.org/6H1okAxEy)

34. Mizzaro, S. & Robertson, S. (2007). HITS hits TREC: exploring IR evaluation results with network analysis. In Wessel Kraaij, Arjen P. de Vries, Charles L. A. Clarke, Norbert Fuhr, Noriko Kando (Eds.). *Proceedings of the 30th Annual International ACM SIGIR Conference on Research and Development in Information Retrieval, Amsterdam, The Netherlands, July 23-27, 2007.* (pp. 479-486). New York, NY: ACM Press. Retrieved 5 May, 2013 from http://www.soi.city.ac.uk/~ser/papers/MizzaroRobertsonSIGIR07.pdf Archived by WebCite® at http://www.webcitation.org/6H1oEKke1)

35. Robertson, S. (2008). On the history of evaluation in IR. *Journal of Information Science,* **34**(4), 439-456

36. Robertson, S.E. & Hancock-Beaulieu, M. (1992). On the evaluation of information retrieval systems. *Information Processing and Management,* **28**(4), 457-466

37. Sanderson, M. (2010). Test collection based evaluation of information retrieval systems. *Foundations and Trends in Information Retrieval,* **4**(4), 247-375. Retrieved 5 May, 2013 from http://www.seg.rmit.edu.au/mark/publications/my_papers/FnTIR.pdf Archived by WebCite® at http://www.webcitation.org/6H1oISFSP)

38. Sanderson, M. & Braschler, M. (2009). Best practices for test collection creation and information retrieval system evaluation. Pisa, Italy: TrebleCLEF. (TrebleCLEF technical report). Retrieved 5 May, 2013 from http://www.seg.rmit.edu.au/mark/publications/

my_papers/T-CLEF-test-collection-report.pdf Archived by WebCite® at http://www.webcitation.org/6H1oLiTdt)

39. Sanderson, M., Paramita, M., Clough, P. & Kanoulas, E. (2010). Do user preferences and evaluation measures line up? In Fabio Crestani, Stéphane Marchand-Maillet, Hsin-Hsi Chen, Efthimis N. Efthimiadis, Jacques Savoy (Eds.). *Proceedings of the 33rd Annual International ACM SIGIR Conference on Research and Development in Information Retrieval, Geneva, Switzerland, July 19-23, 2010.*. (pp. 555-562). New York, NY: ACM Press. Retrieved 5 May, 2013 from http://ir.shef.ac.uk/cloughie/papers/p555-sanderson.pdf Archived by WebCite® at http://www.webcitation.org/6H1oSVAAf)

40. Saracevic, T. (1995). Evaluation of evaluation in information retrieval. In Edward A. Fox, Peter Ingwersen, Raya Fidel (Eds.). *Proceedings of 18th Annual International ACM SIGIR Conference on Research and Development in Information Retrieval, Seattle, Washington, USA, July 9-13, 1995*. (pp. 138-146). New York, NY: ACM Press. Retrieved 5 May, 2013 from http://comminfo.rutgers.edu/~muresan/IR/Docs/Articles/sigirSaracevic1995.pdf Archived by WebCite® at http://www.webcitation.org/6H1oVU1EI)

41. Schamber, L. (1994). Relevance and information behavior. *Annual Review of Information Science and Technology*,**29**, 3-48

42. Smucker, M.D. & Clarke, C.L.A. (2012). Time-based calibration of effectiveness measures. In William R. Hersh, Jamie Callan, Yoelle Maarek, Mark Sanderson (Eds.). *Proceedings of the 35th International ACM SIGIR Conference on Research and Development in Information Retrieval, Portland, OR, USA.*. (pp. 95-104). New York, NY: ACM Press. Retrieved 5 May, 2013 from http://www.mansci.uwaterloo.ca/~msmucker/publications/smucker-clarke-sigir2012.pdf Archived by WebCite® at http://www.webcitation.org/6H1oa1knA)

43. Tague-Sutcliffe, J.M. (1996). Some perspectives on the evaluation of information retrieval systems. *Journal of the American Society for Information Science*, **47**(1), 1-3. Retrieved 5 May, 2013 from

http://comminfo.rutgers.edu/~muresan/IR/Docs/Articles/jasisTagueSutcliffe1996.pdf Archived by WebCite® at http:// www.webcitation.org/6H1oe6O3B)

44. Taube, M. (1956). Cost as the measure of efficiency of storage and retrieval systems. In Mortimer Taube, *Studies in Coordinate Indexing* Volume 3. (pp. 18-33). Washington, DC: Documentation Inc.

45. Turpin, A. & Hersh, W. (2001). Why batch and user evaluations do not give the same results. In W. Bruce Croft, David J. Harper, Donald H. Kraft, Justin Zobel (Eds.). *Proceedings of the 24th Annual International ACM SIGIR Conference on Research and Development in Information Retrieval, September 9-13, 2001, New Orleans, Louisiana, USA*. (pp. 225-231). New York, NY: ACM Press.

46. van Rijsbergen, C.J. (1979). Information retrieval. (2nd ed.) London: Butterworths. Retrieved 5 May, 2013 from http://www.dcs.gla.ac.uk/Keith/Preface.html Archived by WebCite® at http://www.webcitation.org/6H1ool2uv)

47. Voorhees, E. M. (1998). Variations in relevance judgments and the measurement of retrieval effectiveness. In Hans-Peter Frei, Donna Harman, Peter Schäuble, Ross Wilkinson (Eds.). *Proceedings of the 21st Annual International ACM SIGIR Conference on Research and Development in Information Retrieval, August 18-22, 1996, Zurich, Switzerland.* (pp. 315-323). New York, NY: ACM Press. Retrieved 5 May, 2013 from http://nlp.cs.swarthmore.edu/~richardw/ papers/vorhees1999-variations.pdf Archived by WebCite® at http://www.webcitation.org/6H1otD9Jl)

48. Voorhees, E.M. (2000). Variations in relevance judgments and the measurement of retrieval effectiveness.*Information Processing & Management,* **36**(5), 697-716

49. Voorhees, E.M. (2002). The philosophy of information retrieval evaluation. In *Evaluation of cross-language information retrieval systems: 2nd workshop of the Cross-Language Evaluation Forum*

(CLEF 2001). (pp. 355-370). Berlin: Springer. (Lecture Notes in Computer Science 2406).

50. Voorhees, E.M. & Harman, D.K. (2005). *TREC: experiments and evaluation in information retrieval.* Cambridge, MA: MIT Press.

51. Yilmaz, E., Shokouhi, M., Craswell, N. & Robertson, S. (2010). Expected browsing utility for web search evaluation. In *Proceedings of the 19th ACM International Conference on Information and Knowledge Management.* (pp. 1561-1564). New York, NY: ACM Press.

52. Zhai, C., Cohen, W.W. & Lafferty, J. (2003). Beyond independent relevance: methods and evaluation metrics for subtopic retrieval. In *Proceedings of the 26th Annual International ACM SIGIR Conference on Research and Development in Information Retrieval, July 28 - August 1, 2003, Toronto, Canada.*. (pp. 10-17). New York, NY: ACM Press.

53. Zhang, J. & Kamps, J. (2010). A search log-based approach to evaluation. In *Proceedings of the 14th European Conference on Research and Advanced Technology for Digital Libraries.* (pp. 248-260). Berlin: Springer. (Lecture Notes in Computer Science, 6273)

54. Zobel, J. (1998). How reliable are the results of large-scale information retrieval experiments? In *Proceedings of the 21st Annual International ACM SIGIR Conference on Research and Development in Information Retrieval, August 24-28 1998, Melbourne, Australia.* (pp. 307-314). New York, NY: ACM Press. Retrieved 5 May, 2013 from http://goanna.cs.rmit.edu.au/~jz/fulltext/sigir98.pdf Archived by WebCite® at http://www.webcitation.org/6H1p6aqJh)

6

TASK DIMENSIONS OF USER EVALUATIONS OF INFORMATION RETRIEVAL SYSTEMS

F.C.Johnson, J.R. Griffiths and R.J. Hartley

Department of Information and Communications, Manchester Metropolitan University, Manchester, UK

ABSTRACT

This paper reports on the evaluation of three search engines using a variety of user-centred evaluation measures grouped into four criteria of retrieval system performance. This exploratory study of users' evaluations of search engines took as its premise that user system success indicators will derive from the retrieval task the system supports (in its objective to facilitate search). This resulted in the definition of user evaluation as a multidimensional construct which provides a framework to link evaluations to system features in defined user contexts. Our findings indicate that users' evaluations across the engines will vary, and the dimensional approach to evaluation suggests the possible impact of system features. Further analysis suggests a moderating effect on the strength of the evaluation by a characterization of the user and/or query context. The development of this approach to user evaluation may contribute towards a better understanding of system feature and contextual impact on user evaluations of retrieval systems.

1. INTRODUCTION

This paper presents the development of a framework for the evaluation of Information Retrieval (IR) systems, focusing on Internet search engines from a user perspective. The traditional measures for evaluation based on the relevancy of the retrieved output may only be a partial match of users' objectives and of the systems' objectives. The user's judgement of the success of the system may be influenced by factors other than the recall and precision of the output. Such factors are likely to be related to the degree to which the system meets *its* objective to facilitate and maximize the users' search. Usability measures are typically used for the evaluation of an interactive retrieval system and allow some investigation of the impact of system features on users' search behaviour and, in turn, system performance. An alternative approach is to substitute (action based) usability measures for users' assessment of the system on selected variables. Towards this end we posit that users' evaluation of a retrieval system is a multidimensional construct based on the user information searching process which the system seeks to support. In this paper we investigate the development of this criterion on which users might judge the success of a retrieval system. A small scale evaluation of three search engines was carried out using groups of possible success indicators and the interrelations between these variables were examined to suggest those which appear to be determinants of users' judgment of system success (with information retrieval). This initial investigation provides a basis on which we can speculate the value of dimensional user evaluations of system success defined in terms of system suitability for the user task. It is proposed, for further research, that this user-centred approach can provide a framework in which user evaluations on the dimensions can relate to relevant system and situational impacts. If users' evaluations, determined by the search process, are dependent on system features then we can expect evaluation to vary across systems. Further, if a moderating effect of the user query context can be observed then we may be closer to understanding variations in users' evaluations of the same system.

2. SEARCH ENGINE EVALUATION AND DEVELOPMENT

Harter and Hert (1997) in their comprehensive review of evaluation of IR systems define evaluation partly as a process by which system effectiveness is assessed in terms of the degree to which its goals and objectives are accomplished. Modern interactive retrieval engines aim to support the user in the provision of features and functionality which help maximize their search for the retrieval of relevant and pertinent results. The user and their search as an integral component of the system under investigation has been the focus of ongoing developments in evaluation methodologies for interactive retrieval systems (for example, Robertson and Hancock-Beaulieu, 1992 and Harman, 2000. As Dunlop (2000: 1270), reflecting on MIRA, a working group to advance novel evaluation methods for IR applications, states:

the challenge for interactive evaluation in IR is to connect the two types of evaluation: engine performance, and suitability for end-users.

Traditional recall and precision measures of engine performance are based on the concept of relevance, that for a given query there is a set of documents which match the subject matter of that query. Relative recall and precision measures have been used for the evaluation of search engine performance but comparison across these studies is difficult. Different scales for the relevance judgements have been used (Chu and Rosenthal, 1996; Ding and Marchionini, 1996; Leighton and Srivastava, 1999) and/or the results are drawn from different query sets (Gauch and Wang, 1996; Tomaiuolo and Packer, 1996; Back, 2000). It was, however, the long standing criticism of the very basis of these performance measures, binary relevance judgments, which was highlighted as a major criticism of the evaluations of web search engines participating in the Web Special Interest track of the large scale testing environment of the Text Retrieval Conferences (TREC-8) (Hawking *et al.*, 1999; Hawking *et al.*, 2001).

Sherman's report of the Infonortics 5th search engine meeting (2000) states that the relevancy of the system output is a poor match of these systems' objectives which he suggests include getting information from users, providing browsing categories, promoting popular sites and speed of results. Those, it seems, which assist a user to express a query, navigate through the collection, or quickly get to the requested information. This trend in the development of search assistance features supporting casual users has been consistently noted (Sullivan, 2000; Feldman, 1998, 1999; Wiggens and Matthews, 1998). Search and retrieval features are touted as maximising user capabilities in manipulating the information searching process. Statistical probablistic retrieval systems allow natural language or, more precisely, unstructured queries which may support the user's task of query formulation; concept processing of a search statement may determine the probable intent of a search; relevance feedback can assist users in modifying a query; and, use of on and off the page indicators to rank the retrieved items may assist users in judging the hit list, as do visualisation techniques, for example, which may be used to provide a view of a subset of the collection.

The increasingly interactive nature of system design motivates approaches to evaluation which accommodate the user and the process of interaction. Front-end features will affect users' interactions which in turn will partly determine the effective performance of the back-end system. The requirements which can be derived from this interaction of system features, user and query are articulated well in Belkin *et al.*, as highlighted in Harter and Hert, 1997,: 26

if we are going to be serious about evaluating effectiveness of interactive IR, we [sic. need to] develop measures based upon the search process itself and upon the task which has lead the searchers to engage in the IR situation.

The rise of interactivity was acknowledged from TREC-3 onwards with the introduction of a special interactive track (Beaulieu *et al.*, 1996) and its goal to investigate the process as well as the outcome in interactive searching (Hersh and Over, 2001) Various usability measures of the user-system performance, such as number of tasks completed, number of query terms entered, number of commands used, number of cycles or query reformulations, number of errors and time taken, were thus derived by consideration of the users' actions or behaviour in carrying out a search task. As such, usability measures may provide indicators of the impact specific features of a retrieval system have on searcher behaviour. Voorhees and Garofolo, (2000) itemise studies which have investigated features such as the effects of visualization techniques, and different styles of interaction, while several studies have focused on the system feature of relevance feedback in interactive IR. Belkin *et al.* (2001) looked at query reformulation, and White *et al.* (2001) compared implicit and explicit feedback. In the context of TREC-8 interactive track, Fowkes and Beaulieu (2000) examined searching behaviour, related to the query formulation and reformulation stages of an interactive search process, with a relevance feedback system. Of particular note in this study was the moderating effect of the user query context, where different query expansion techniques were found to be suited to the degree of query complexity.

Others substitute usability measures for satisfaction measures and ask users about their satisfaction with or judgment of the general performance of the system and possibly its specific features (or more widely, the interface). Su (1992, 1998) identified twenty user measures of system success which were grouped into the evaluation dimensions of relevance, efficiency, utility and user satisfaction. These were correlated to the users' rating for overall system success to determine 'value of search results as a whole' to be the best single user measure of success. The pursuit to find the best determinant of users' system success rating is a worthy one which would reduce considerable cost and effort in evaluation. Yet its application would mask the potentially complex judgment resulting from the many factors with which a user engages in interacting with the system and on which the user may draw in assessing the system. The crux of the issue being that while a user evaluation (expressed in a single construct) is more easily obtained, it is at the expense of knowing why. From the system developers' perspective it may be of value to have greater insight into user evaluations and the subtle balance of the interrelations between the various indicators of a user judgment of system success. A single system, for example, may be rated as providing excellent results but requiring considerable user time and effort, thus high on effectiveness but low in efficiency or indeed visa versa. Hildreth (2001), for example, found that 'perceived ease of use' of an OPAC was related to 'user satisfaction with the results'. Our concern with what might be the determinants of a user judgment such as 'ease of use' leads to consideration of a more detailed system evaluation. For example, a system's layout and representation of the retrieved items may contribute to a user perception of ease of use helping the user to quickly and easily make relevancy judgments. It would seem reasonable to speculate that a system could score high in the user's judgment of this aspect but, in fact, score low on user satisfaction with the search results.

2.1 Dimensions and determinants of user evaluation.

A system evaluation based on various success indicators requires some basis on which the interrelations between these factors can be sought and understood. This requirement is perhaps demonstrated by the fact that previous studies have been unable to establish a consistent relation between user satisfaction and the recall and precision of the search results (Sandore, 1990; Su, 1992; Gluck, 1996; and Saracevic and Kantor, 1988). The reason for this may be that it is an erroneous assumption that there ought to be a relation between user and system measures of performance. It would certainly be dependent on the individual and their situation. The framework we seek to develop is thus one which relates and groups various user indicators of success into dimensions of the users' overall judgment in order that meaningful relations can be sought in the evaluation. The remainder of this paper describes our preliminary investigation into the defining function of the retrieval task dimensions on user evaluation indicators and measures. Ideally, user evaluation based on what the user is doing and the system is supporting will link to system features in well defined search contexts. A further requirement for our feasibility study is thus to identify possible user/query contexts which may moderate users' judgments of the system. Users with different demands on the system are likely to vary accordingly in their assessment based on the system and its features. Succinctly put, our feasibility study set out to identify the possible dimensions and indicators of user evaluations in a framework in which evaluation is dependent on system features and moderated in a user/query context.

2.2 The IR task process

Several models of information searching have been suggested which may be used to identify the user success factors in system evaluation. Information retrieval, for example, may be viewed as an exploratory process, a view which led Brajnik (1999) to derive evaluation statements for criteria such as system flexibility. The approach we explored focused on the more mechanical process models to demarcate the tangible (system dependent) information retrieval activities. Such models of the basics of information searching are standard, for example in Salton (1989) as highlighted in Baeza-Yates (1999: 262), and from which the following interacting steps are defined:

- formulation and submission of a query,
- examination of the results, *with a*
- possible feedback loop to re-formulate the query, *and*
- integration of search results and evaluation of the whole search.

Each step provides some statement of user-requirement, what the goal-directed user is trying to do with the system. Each step in the process thus represents a dimension which defines the variables and measures on which a user may evaluate a system's success in supporting information retrieval.

Table 1 groups these indicators of system success under each of the dimensions, Query formulation, Query reformulation, Examination of result, and Evaluation of search results and search as a whole. The majority of which came from existing (and generally accepted) measures of retrieval system performance (for example as listed in Su (1992) and which form the broad criteria of effectiveness, utility and efficiency. The main effect of using the lower level task dimensions is the decomposition of interaction by the three task process steps.

Table 1. Indicators of user judgment of system success grouped by task dimensions

Effectiveness	Utility	Efficiency
Evaluation of results Satisfaction with precision Satisfaction with ranking	**Evaluation of search and results** Value of search results Satisfaction with results Resolution of the problem Rate value of participation Quality of results	**Evaluation of search as a whole** Search session time Response time
Interaction		
Query formulation Satisfaction with query input	**Query reformulation** Satisfaction with query modification Satisfaction with query visualisation	**Examine results** Satisfaction with visualisation of item Satisfaction with manipulation of output

2.3 Measures of search results

Users' judgment of the success of the search results and the search as a whole may be based on the criterion of effectiveness manifest in the system output, the retrieved set. Traditional measures of retrieval effectiveness are based on the notion of relevance, and evaluation of the system will be partially dependent on the ability of the system to meet its basic function to retrieve only relevant documents. We used Su's user measures of user satisfaction with precision, and ranking. Further indicators of a user success judgment of the search may be based on factors, other than relevance. Measures of utility focus on the actual usefulness or value of the retrieved items to the individual information seeker (Saracevic *et al.*, 1988). Cleverdon (1991) argued (with Cooper, 1973, who put forward a straight utility-theory single measure) that retrieval effectiveness measures of recall and precision should be used in combination with (and possibly related to) these more user-oriented measures which are based on factors as subjective satisfaction statements, search costs, and time spent. Indeed, various factors may bear on users' judgements of overall satisfaction with the value of the search results. For example, users may be influenced by the extent to which information quality can be assumed based on the source; the extent to which the information is accurate or correct; and, the extent to which the information is at the right level to meet user need. Yet, Saracevic (and Su, 1998: 558) report that standard utility measures do not exist. Saracevic used the following evaluative statements:

- How much time spent reviewing abstracts;
- Assign a cost value to usefulness of results;
- What contribution this information made to resolution of problem that motivated your question;
- Overall how satisfied with results.

Su found that a measure of utility **value of the search results as a whole** correlated most strongly with users' overall judgement of system success and thus proposed it to be a best single measure of a system. We used as a measure of utility *Satisfaction with results* and *resolution of the problem* (derived from Saracevic) and *value of search results as a whole, value of participation* and *quality of results* (from Su).

2.4 Measures of user-system interaction

In the process of getting the search results the user is involved in the steps of query formulation, re-formulation and examination of the results. These comprise broad categories of user interaction with the system, and the specific actions subsumed relate to how the user interacts and manipulates or commands the system to retrieve the required information. In the absence of usability measures, interaction will be largely determined by satisfaction measures alone. As Belkin and Vickery (1985: 194) point out satisfaction is a concept intended to capture an overall judgement based on user reaction to the system thus extending the range of factors relevant to the evaluation. Our approach which has interaction as a broad category of the three interactive task dimensions breaks down the concept to measures based on these relevant factors, grouped in Table 1 and explained as follows. **Query formulation**, on consulting an information retrieval system, user reaction to the system may be influenced by the perceived ease of expressing the query. The user may be influenced in this judgement by, for example, the availability of different search methods, such as natural language searching or power search to specify a search topic. A measure of *user satisfaction with query input* thus is defined in terms of the perceived ease in the expression or specification of the query. **Search reformulation**, users may be influenced by any assistance or feedback received for formulating or modifying the query, such as the system suggesting query terms from a thesaurus or offering 'more like this' feedback options. Another form of feedback lies with the query visualization, the assistance provided in understanding the impact of a query. An example is the use of folders to categorise search results which may suggest to the user different perspectives of the topic which is useful in refining the search. The measures of *user satisfaction with query modification* and *satisfaction with query visualization* are thus defined in terms of system suggesting search terms or facilitating query by example, and in terms of understanding the impact of the query respectively. **Examining the results**, on receiving results the user will be involved in some process of interpreting the results in the given frame of the information need and would want to see quickly and easily an item's topic or meaning and why it was retrieved. Summary representation features for visualising the 'aboutness'of an item might support a user in this task, e.g., in highlighting query terms, showing category labels and a clear and organised layout. Thus we defined the *measure of satisfaction with visualisation of item representation* and*manipulation of the output* (e.g., summary display features (category labels), sort by).

2.5 Measures of search efficiency

The final stage in the retrieval task is evaluation of the search as a whole and relates to the criterion of efficiency. Boyce *et al.* (1994: 241) highlight the difference between effectiveness and efficiency thus:

an effectiveness measure is one which measures the general ability of a system to achieve its goals. It is thus user oriented. An efficiency measure considers units of goods or services provided per unit of resources provided.

Dong and Su (1997: 79) state that response time is becoming a very important issue for many users. If users want to retrieve information as quickly as possible, this may in part equate to the efficiency of the system and the judgement of which will affect user evaluation of the system success as a whole. Thus whilst efficiency seems hard to define, these studies and others (such as, Stobart and Kerridge, 1996; Nahl, 1998) seem corroborate on the importance of speed of response as an indicator of efficiency.

2.6 Multidimensional framework for search engine evaluation

In our feasibility study these indicators of satisfaction were used to elicit a user judgment of the success of three search engines. Our aim was not to obtain an evaluation of these engines as such but rather to confirm or refute our premise that a dimensional user evaluation can result in meaningful relations sought in the user judgments made across and within the systems. This can be stated in three propositions which were explored to varying degrees afforded by the scope of the study.

Proposition One states that the user judgment of system success is a response to how well the system has supported the retrieval task and, as such, is a multidimensional construct. The nature of user evaluations was explored as follows:

- User success ratings assigned on the four criteria were correlated with an overall success judgment to find which, if any, appears be the most important factor in defining user judgment of the system.
- Users ratings on the measures were correlated with the overall success ratings for each associated criterion to find which, if any, contributed most strongly to the user's overall rating of a criterion.
- User derived reasons for attributing satisfaction ratings, overall and on each criterion, were collected using open-ended questions and analysed to suggest, or otherwise, our measures as those which users themselves base an evaluation of system success.

Proposition Two states that user ratings will vary across the systems reflecting the support of system features to the retrieval task. In the scope of this feasibility study we do not claim to test this, but evidence of it was sought in the finding that user evaluations varied across the engines.

- User ratings on each of the measures were compared across the search engines to find which engine, if any, received notably higher/lower ratings. Some speculation was made as to the possible impact of system features.

Proposition Three states that user evaluations of the systems will be moderated by the context of the users' information query making different demands on the system. Again this was not tested but evidence was sought of contexts leading to the systems receiving different evaluations.

- Four task identifiers were analysed against the overall user ratings and the four evaluation criteria across all four search engines to ascertain if a moderating effect of context was obtained.

2.7 Implementation

Twenty-three participants were recruited from second year students of the Department of Information and Communications, MMU. A short briefing was given a few days prior to their search session to explain the project and to present the Information Need Characteristic Questionnaire which was to be completed before the search session. No restrictions were placed on the type of information sought or the purpose for which it was intended, but the questionnaire did capture a characterisation of the search context in terms of following parameters used in Saracevic (1988): a) problem definition (*on a scale from 1-5, would you describe your problem as weakly defined or clearly defined?*); b) intent (*on a scale from 1-5, would you say that your use of this information will be open to many avenues, or for a specifically defined purpose*); c) amount of prior knowledge (*on a scale from 1-5, how would you rank the amount of knowledge you possess in relation to the problem which motivated the request?*); and d) expectation (*on a scale from 1-5 how would you rank the probability that information about the problem which motivated this research question will be found in the literature?*). Participants were then asked to conduct the search using as many reformulations as required and to search for as long as they would under normal conditions. Each was required to search the three engines, Excite, NorthernLight, and HotBot the order of which was varied to remove learning curve effect. These engines were selected on the basis that each had at least one discernable feature so that each search would present a unique search experience and the ability to distinguish the engines.

Following each search participants were required to respond on a likert type scale to questions relating to each of the user satisfaction variables indicated in**Table 1** as defining each of the evaluation criterion. In addition users were asked to provide an overall success rating of the engine with respect to each criterion, i.e., effectiveness, efficiency, utility and interaction. This was to allow the testing of Proposition one as stated above towards the identification of the measures which when validated could form the basis to comprise a judgment for each criterion.

To measure system effectiveness users were asked to rate on a three point scale the degree of relevance of each item retrieved, leaving it open as to how many individual items were assessed. The searchers were provided with definitions of relevant, partially relevant and non-relevant which had their origins in the ISILT (Information Science Index language test) project undertaken by Keen and Digger in the early 1970s (Keen, 1973 ,) and which have been used in various tests since that time. Since the searches were carried out on three different engines it is highly likely that identical items would be retrieved with a possibility that 'already seen' items are consciously or unconsciously judged to be less relevant the second or third time around. This is partly resolved in the varying order of the engines to which searchers were assigned. In this instance, however the impact was not considered to be great as the aim was not to evaluate the performance of the individual engines per se but to obtain a user's expression of satisfaction with precision across the engines' results. Participants were then asked to rate on a five point scale their **satisfaction with the precision** of the search results. An overall rating of effectiveness was obtained by the users' assessment of the ***overall success*** of the search engine in retrieving items relevant to the information problem or purpose on a five point scale. To measure utility users were asked to rate on a five point scale the **worth of their participation**, with respect to the resulting information; the contribution the information made to **the resolution of the problem; satisfaction with results; the quality of the results**; and the **value of the search results as a whole**. Participants were asked to rate on a five point scale the ***overall success*** of the search engine in terms of the actual usefulness of the items retrieved. To measure interaction the users were asked to rate on a five point scale satisfaction with **query input, query modification, query**

visualisation, manipulation of output and**satisfaction with visualisation of representation** of item. To measure efficiency participants were required to record the **search session time** and to rate on a five point scale the ***overall success*** of the search engine in retrieving items efficiently.

2.8 Analysis and interpretation

Our primary aim, expressed in proposition one, was to explore the multidimensional nature of users' evaluation of a system. To begin to understand how users might themselves evaluate these systems we correlated users' overall success rating with the ratings assigned on the four criteria to suggest which, if any, appears be the most important or contributory factor to users' overall judgement of system success. The results are shown in Table 2: where a moderate correlation is defined as greater than 0.4 and less than 0.7 and a strong correlation as between 0.7 and 0.9.

Table 2. Global and engine level - Overall success rating correlated against the four criteria

	Spearman's rank correlation coefficient **=strong *=moderate strength correlation			
Criterion	**Global**	**Excite**	**NorthernLight**	**Hotbot**
Effectiveness	.759**	.779**	.795**	.729**
Efficiency	.817**	.843**	.908**	.741**
Utility	.710**	.362	.930**	.806**
Interaction	.592*	.511*	.660*	.580*

The strength of the correlation ratings indicates that users' overall success rating of the system appears to be multidimensional. The Efficiency criterion held the strongest correlation with users' overall rating (.817) and Interaction the weakest (.592) which is in accordance with other studies which suggest ease of use and response time to be strong determinants of satisfaction. This result is fairly consistent with the analysis done at the level of the individual search engine, although we note the strong correlation with utility on NorthernLight and HotBot which was not found with Excite.

2.9 Indicators of the dimensions

Correlations were also taken of the measures within each criterion with users' overall rating. Strong correlations may suggest single measures as the best indicator of users' evaluation of the criterion. Results at the global level, across the three engines, suggest that 'satisfaction with precision' has the strongest correlation (.733) with the judgement of the systems' effectiveness. All the measures of utility held a strong correlation with its overall judgement, the strongest measure being 'satisfaction with results' (.755). The measure 'rate worth of participation' had a negative correlation which indicates that as utility rises the value of participation decreases. This may bring into question users' interpretation of value of participation. We can only speculate but it was possibly mistaken to refer to amount of user effort exerted. Interestingly, the measure of 'time taken to search' held a negligible correlation (.062) with the judgement of the systems' efficiency. The judgement of the satisfaction with systems' interaction held a correlation of moderate strength (.506) with the measure 'satisfaction with query visualization'. This is followed by 'satisfaction with facility to input query' (.486), 'satisfaction with visualization of item representation' to ease understanding of item/s from the hitlist (.452), 'ability to modify query' (.437), and 'ability to manipulate output' (.132), this being a very weak correlation. At the level of individual search engines, the judgement of interaction on both Excite and HotBot held the strongest correlation with satisfaction with query visualization, whereas NorthernLight held the strongest correlation with the ability to modify the query. The low correlations of the measure of efficiency and the negative correlation of the measure of value of participation are considered in the concluding section of this paper.

User derived reasons for attributing satisfaction ratings, overall and on each criterion, were collected using open-ended questions. These were analysed to suggest the extent to which our selected measures compared to those which users themselves might base an evaluation of system success. Some 250 comments were collected and, in the main, simply confirm the users' understanding of our intended interpretation of system effectiveness and utility. The selected, but representative, comments in **Table 3** show the user-derived reasons for assigning ratings of success of effectiveness seem to confirm the finding that user satisfaction with precision held the strongest correlation with user ratings of effectiveness. The utility measures all held strong correlations and the user-derived reasons also indicate that these were measures which users may themselves use. Further analysis would be required to ascertain if in fact these measures were simply variations of the same measure 'satisfaction with results'.

Table 3. User derived reasons for assigning success ratings on the four criteria.

Effectiveness 'Information retrieved was extremely relevant to my needs.' 'Most items retrieved appear to have some relevance.' 'Too much irrelevant information.' **Efficiency** 'Ease of use.' 'Had to redefine search twice.' 'The search terms were attempting to pin down a concept that was hard to verbalize [or] encapsulate.' 'Would become "extremely efficient" as the user becomes more adept with search terminology phrasing and when an "advanced search" would be more appropriate. 'Very quick. only had to search once.' 'One search term located all items that were of some relevance.' 'Needed to define search better.' 'Minimum effort, but results not good.' 'Search engine seemed efficient enough. but the search term was unusual. I think with a more concrete search term the SE would have performed well.'	**Utility** 'Those items were found useful.' 'I have gained further info[rmation] on the subject I was search[ing] for.' 'Current info[rmation] was located.' **Interaction** 'I changed the query once and it was helpful.' 'The SE easily allowed the query to be modified.' 'Found it hard to refine search.' 'The query was easy to change but yielded no better results.' 'Good options to change query.' 'Refining the search was hard. I couldn't think of any new queries and the SE didn't offer any help trying to narrow down search queries, like the SE I usually use.' 'Could lead to different routes of enquiry from the initial search term.'

The correlations found with the interaction measures were relatively low with satisfaction with query visualization holding the strongest correlation. The user derived reasons, however, indicate that there was perhaps some expectation that the system would provide some assistance in modifying the query and that this would impact on evaluation of system interaction. The efficiency criterion held the strongest correlation with an overall judgment of success, yet the measure of search time held a low correlation as a measure of efficiency. Interestingly, the user-derived reasons for assigning ratings of success on this criterion suggest that users relate efficiency, not to time taken but to something equating to the amount of user effort required to conduct a search. The real worth of this investigation is that while we may suggest that user evaluation is multidimensional and there are some obvious candidate variables on which to base the measure of these evaluation dimensions, considerable research is still needed to properly ascertain and validate user measures. We still need to understand how users themselves evaluate a system and to what these measures relate.

2.10 Feature impact

Our second proposition is that system characteristics will affect user evaluations on the task dimensions which have defined the evaluation. Users' ratings varied across the search engines and, within the scope of a small scale study, this was viewed as indicative that user evaluations are not random but an elicited response reflecting the support of the system to the users' task. As has been noted, strong correlations with users' overall rating and utility was found on NorthernLight and HotBot, in contrast to the very weak correlation found on Excite. The marked difference in the strength of correlation found between the systems is interesting but only in that it **suggests** that users' overall judgement may be more strongly associated with a judgement made on a particular criterion depending on the system. In a full scale evaluation study with a far larger sample more insight and interpretation could be possible from an analysis of the central tendency on the rating scales for the individual measures. For example, taking the strongest correlation with users' overall rating and interaction on NorthernLight (see Table 2), it was found that 77% of users of this system rated the ability to modify the query whereas globally and on the other two systems the measure of user satisfaction with query visualization had the strongest correlation . Again this may indicate the influence of a system feature on the users' judgement, but this cannot be substantiated without a much larger investigation.

2.11 Impact of query context

Our final proposition was that users' evaluations of the system may be moderated by some contextual characterisation of the user and information query. To obtain some indication of the support for this proposition we analysed the user/query context where a system received high/low ratings by correlating the four task identifiers (task defined, task purpose, task knowledge and task probability) against the overall satisfaction rating and the four criteria. Again we stress the exploratory nature of this exercise and that a greater sample would be required in an evaluation situation to support any analysis at this level.

Globally, across the three engines, moderate strength correlations indicated that as *task definition* increases so does overall rating of system success (.407), and satisfaction with effectiveness (.418) and efficiency (.482). The correlations between *task definition* and utility (.307) and interaction (.221) were weak. Weak or very weak correlations were obtained between *task purpose, task knowledge*, and *task probability* and the overall success rating and the four criteria. The suggestion that a system receives a higher rating of effectiveness and efficiency when the user has a well-defined task is not surprising. It would be reasonable to assume that in such a context the information seeker will have a fairly good idea of the search requirements to obtain good results. Indeed, the effect of the moderating context will be of more interest if notable variations can be found between the systems evaluated.

Indicative of such a finding is with the system NorthernLight. Whilst weak correlations were found globally for *task purpose*, when based on the data obtained for NorthernLight moderate strength correlations were found with Efficiency (.636) and Utility (.577). The comparison is with the weak correlation found on Excite data with Efficiency (.161) and Utility (.002). It would not be surprising if a correlation was found for intent of task purpose and utility across all three engines: a broad query, open to many avenues, could lead to a high rating of the utility of the results. That such a finding is strongly held only on one engine could lead to speculation that a feature of the engine leads to results which better support a broad query. For example, NorthernLight boosts features related to the visual organisation and representation of the search results which may better support the user with a broad query. Indeed users of the system expressed a high level of satisfaction with item visualization, in terms of understanding the content of the hit list, and query visualization in terms of understanding the impact of the query on the results obtained. A larger scale study would be necessary to ascertain the significance of this rating when compared to others.

3. CONCLUSIONS

Our preliminary investigation into user evaluations of internet search engines would seem to indicate that these are determined by many factors which together may represent dimensions of some overall user judgment of the system. To explore the value of a framework for evaluation based on multi-dimensions we defined and grouped possible indicators of success on the search task process in which the user is engaged. This would seem to be a reasonable approach to the substitution of usability measures for the evaluation of interactive system objectives to maximize search for information retrieval. The success indicators of user satisfaction were drawn from existing measures or by their definition provided by the consideration of the system feature objectives. However, there is a need to ascertain from users themselves what these success indicators might be and to validate these as relating to task dimensions. This is clear, not least, from our finding of some discrepancy in how users might define system efficiency with user effort being a key influence on the user judgment of the system. Further research is thus required to develop the multidimensional construct of user evaluation as possibly a function of the system support for the user retrieval task process.

Ultimately our aim in developing the framework is to have user evaluations link to system features, thus allowing a system to score high in particular aspects but not necessarily in all aspects. In this study, we can only speculate with caution that a system feature of query modification contributed to the users' evaluations, and that there was some observed effect of task purpose as a moderating variable in one system. While this tempts comment, we can only speculate that this was attributed to a system feature which better supported a particular query context. Only in a full-scale evaluation could this be tested using appropriate statistical techniques, such as regression analysis, to express user judgment of the system as a function of the task defined indicators of success and to explore, within and across systems, the relationship held among the dependent and moderating variables. Obviously this would be an expensive undertaking, but one which would not only allow variation to be found in users' assessments across systems influenced by system features, but also variations within search dimensions as influenced by user query task contexts.

ACKNOWLEDGEMENTS

This paper reports on results of a feasibility study of user evaluation of internet search engines, Devise (Johnson *et al.*, 2001). Thanks goes to Re:source for its funding of Devise under grant number NRG/085 from January 2000-2001. The full report is available from our website http://www.mmu.ac.uk/h-ss/dic/cerlim/ and can be requested from the thesis and reports lending division of the British Library. This research is being developed in Department of Information & Communications at MMU, and thanks to Sarah Crudge for many interesting discussions on its development towards the registration of her PhD on user evaluations of search engines.

REFERENCES

1. Back, J. (2000). An evaluation of relevancy ranking techniques used by Internet search engines. *Library and information research news,* **24**(77), 30-34.

2. Baeza-Yates,R. and Ribeiro-Neto, B. (1999). *Modern information retrieval* Reading, MA: Addison-Wesley.

3. Belkin, N.J., Cool, C., Kelly, D., Lin, S.J., Park, S.Y., Perez-Carballo, J., Sikora, C. (2001). Iterative exploration, design and evaluation of support for query reformulation in interactive information retrieval." *Information processing and management,* **37**(3), 403-434.

4. Beaulieu, M., Robertson, S., and Rasmussen, E. (1996). Evaluating interactive systems in TREC. *Journal of American Society for Information Science,* **47**(1). 85-94.

5. Belkin, N.J. and Vickery, A. (1985). *Interaction in information systems: a review of research from document retrieval to knowledge-based system*. London: the British Library.

6. Boyce, B.R., Meadow, C.T., and Kraft, D.H. (1994). *The measurement of information science*. Reading, MA: Academic Press.

7. Brajnik, G. (1999). Information seeking as explorative learning, in: Proceedings of the MIRA '99 conference." In: *the electronic workshop in computing series.* [Available at http://www.ewic.org.uk]

8. Brajnik, G. (1999). Information seeking as explorative learning. In: S. W. Draper, M. D. Dunlop, I. Ruthven, and C.J. Van Rijsbergen, *eds., Proceedings of Mira 99: Evaluating Interactive Information Retrieval, Glasgow, April 1999.* British Computer Society, Electronic Workshops in Computing. Retrieved 17 July 2003 from http://www1.bcs.org.uk/DocsRepository/02800/2836/brajnik.pdf

9. Chu, H. and Rosenthal, M."Search engines for the world wide web: A comparative study and evaluation methodology." In: *ASIS '96: Proceedings of the 59th ASIS annual meeting*, 33, pp.127-135. Medford, NJ: Information Today. Retrieved 8 July 2003 from http://www.asis.org/annual-96/ElectronicProceedings/chu.html

10. Cleverdon, Cyril W. (1991). The significance of the Cranfield tests on indexing languages. In: *Proceedings of the 14th International Conference on Research and Development in Information Retrieval (ACM SIGIR '91*), pp. 3-12. New York:ACM Press.

11. Cooper, W.S. (1973). On selecting a measure of retrieval effectiveness. *Journal of the American Society for Information Science*, **24**, 87-100.

12. Ding, W.and Marchionini, G. (1996). A comparative study of Web search service performance. In: Hardin, S. *ed., Proceedings of the 59th Annual Meeting of the American Society for Information Science* Vol 33. (pp. 136-142.) Baltimore, MD: Information Today.

13. Dong, X., and Su, L. (1997). A comparative study of Web search service performance. In: C. Schwartz and M. Rorvig, *eds. Proceedings of the 60th Annual Meeting of the American Society for Information Science.* Vol 34. (pp. 136-142). Medford, NJ: Information Today.

14. Dunlop, M. (2000). Reflections on Mira: interaction evaluation in information retrieval." *Journal of the American Society for Information Science,* **51**(14), 1269-1274.

15. Feldman, S. (1998). Web search services in 1998: trends and challenges. *Searcher,* **6**(6), 29-39. Retrieved 8 July 2003 from www.infotoday.com/searcher/jun98/story2.htm

16. Feldman, S. (1999). Search engines: the 1999 conference. *Information Today,* **16**(6). Retrieved 8 July 2003 from http://www.infotoday.com/IT/jun99/feldman.htm

17. Fowkes, H. and Beaulieu, M. (2000). Interactive searching behaviour: Okapi experiment for Trec-8. *Paper presented at the British Computer Society Information Retrieval Special Group 22nd Annual Colloquium on Information Retrieval Research, Cambridge, 5-7 April.* Retrieved 17 July from http://irsg.eu.org/irsg2000online/papers/fowkes.htm

18. Gauch, S. and Wang, G. (1996). Information fusion with ProFusion. In: H. Maurer, *ed. Proceedings of the World Conference of the Web Society (Webnet '96), San Francisco, CA, Oct 15-19.* Retrieved 8th July from http://www.csbs.utsa.edu:80/info/webnet96/html/155.htm.

19. Gluck, M. (1996). Exploring the relationship between user satisfaction and relevance in information systems. *Information Processing and Management,* **32**(1).11-18.

20. Harman, D. (2000). What we have learned and have not learned from TREC. *Paper presented at the British Computer Society Information Retrieval Special Group 22ndAnnual Colloquium on Information Retrieval Research,* Cambridge, 5-7 April.

21. Harter, Stephen, P. and Hert, C.A. (1997). Evaluation of information retrieval systems: approaches, issues, and methods. *Annual Review of information Science and Technology*, **32**, 3-94.

22. Hawking, D., Craswell, N., Thistlewaite, P., Harman, D. (1999). Results and challenges in web search evaluation. In: *The Eighth International World Wide Web Conference,* Toronto, May 11-14. Retrieved 8 July 2003 from http://www8.org/w8-papers/2c-search-discover/results/results.html

23. Hawking, D., Craswell, N., Bailey, P., Griffiths, K. (2001). Measuring search engine quality. *Information Retrieval,* **4**. 33-59.

24. Hersh, W., Over, P. (2001). TREC –9 Interactive Track Report." In: *The Ninth Text Retrieval Conference (TREC –9*). Gaithersburg, MD: National Institute for Standards and Technology. Retrieved 8 July 2003 from http://trec.nist.gov/pubs/trec9/t9_proceedings.html

25. Hildreth, C.R. (2001). Accounting for users' inflated assessments of on-line catalogue search performance and usefulness: an experimental study. *Information Research,***6**(2). Retrieved 8 July 2003 from http://informationr.net/ir/6-2/paper101.html.

26. Johnson, F. C., Griffiths, J.R., and Hartley, R.J. (2001). *Devise: a framework for the evaluation of Internet search engines.* London: Resource: the Council for Museums, Archives and Libraries. (Library and Information Commission Research Report 100)

27. Keen, E.M. and Digger, J.A. (1972) *Report of an information science index language test.* Aberystwyth: College of Librarianship Wales.

28. Keen, E.M. (1973) The Aberystwyth index language tests. *Journal of Documentation* , 29(1), 1-35.

29. Leighton, H.V. and Srivastava, J. (1999). First 20 precision among World Wide Web search services (search engines). *Journal of the American Society for Information Science,* **50**(10), 870-881.

30. Nahl, D. (1998). Ethnography of novices' first use of Web search engines: affective control in cognitive processing. *Internet Reference Services Quarterly*, **32**(2), 69.

31. Robertson, S.E. and Hancock-Beaulieu, M. (1992). On the evaluation of IR systems. *Information Processing and Management*, **28**(4), 457-466.

32. Salton, G. (1989). *Automatic text processing: the transformation, analysis and retrieval of information by computer*. Reading, MA: Addison-Wesley.

33. Sandore, B. (1990). Online searching: what measure satisfaction? *Library and Information Science Research*, **12**, 33-54.

34. Saracevic, T., Kantor, P., Chamis, A.Y., and Tirvison, D. (1988) A study of information seeking and retrieving. I Background and methodology." *Journal of the American Society for Information Science*, **39**(3), 161-176.

35. Saracevic, T. and Kantor, P. (1988). A study of information seeking and retrieving. II. Users, questions and effectiveness. *Journal of the American Society for Information Science*, **39**(3), 177-196.

36. Sherman, C. (2000). *The FireworksFly* [Available at websearch. about.com/library/weekly/ aa041800b.htm]

37. Sherman, C. (2000). *'Old economy' information retrieval clashes with 'new economy' Web upstarts at the Fifth Annual Search Engine Conference: Conference Report.*Medford, NJ: Information Today. Retrieved 17 July 2003 from http://www.infotoday.com/ newsbreaks/nb000424-2.htm

38. Stobart, S. and Kerridge, S. (1996). An investigation into World Wide Web search engine use from within the UK –preliminary findings *Ariadne*, No. 6. Retrieved 8 July 2003 from http://www.ariadne. ac.uk/issue6/survey/

39. Su, L. (1992). Evaluation measures for interactive information retrieval. *Information Processing and Management*, **28**(4), 503-516.

40. Su, L. (1998). Value of search results as a whole as the best single measure of information retrieval performance. *Information Processing and Management*, **34(**5), 57-579.

41. Sullivan, D. (2000). Web search engine trends and achievements since the 1999 Boston Search Engine meeting, In: *Search Engines Today and the New Frontier: the Fifth Search Engine Meeting.*, Boston, Massachusetts, April 2000. Retrieved 8 July 2003 from http://www.infonortics.com/searchengines/sh00/sullivan_files/frame.htm. [PowerPoint presentation]

42. Tomaiuolo, N.G. and Packer, J.G. (1996). An analysis of Internet search engines: assessment of over 200 search queries. *Computers in Libraries*, **16**(6), 58-62.

43. Voorhees, E. and Garofolo, J. (2000). The TREC spoken document retrieval track. *Bulletin of the American Society for Information Science*, **26**(5). Retrieved 17 July 2003 from http://www.asis.org/ Bulletin/June-00/voorheesgarofolo.html

44. White, R.W., Jose, J.M. and Ruthven, I. (2001). Comparing explicit and implicit feedback techniques for web retrieval: TREC-10 Interactive track report, in:*The tenth Text Retrieval Conference (TREC 2001), Gaithersburg, Maryland, November 13-16.* Gaithersburg, MD: National Institute of Standards and Technology. Retrieved 8 July 2003 from http://trec.nist.gov/pubs/ trec10/ t10_proceedings.html

45. Wiggins, R. and Matthews, J.(1998). Plateaus, peaks and promises: the Infonortics '98 search engine conference. *Searcher*, **6**(6). Retrieved 8 July 2003 from http://www.infotoday.com/ searcher/ jun98/story4.htm

7

SHIFTS IN INFORMATION-SEEKING STRATEGIES IN INFORMATION RETRIEVAL IN THE DIGITAL AGE. A PLANNED-SITUATIONAL MODEL

Hong (Iris) Xie

School of Information Studies, University of Wisconsin-Milwaukee, Milwaukee, Wisconsin, USA

ABSTRACT

Introduction

The digital age brings changes to information retrieval systems, users, information, and the environments in which users interact with systems. That also poses challenges for users to effectively retrieve information to accomplish their tasks and goals.

Argument

This paper presents the planned-situational model that illustrates how plans and situations co-determine users' selection of information-seeking strategies and shifts in information-seeking strategies in their information retrieval process in the digital age. The major components of the model include: levels of user goals, levels of tasks and their dimensions, personal information infrastructure, the social-organizational context, information retrieval systems, plans and their dimensions, situations and their dimensions, and information-seeking strategies consisting of interactive intentions and retrieval tactics.

Elaboration

This model is developed based on existing general interactive information retrieval models, influential user-oriented approaches in information seeking and retrieval, specific models related to interactive information retrieval, as well as empirical interactive information retrieval studies in different types of digital environments including online, public-access, catalogueues, online databases, Web search engines, and digital libraries.

Conclusion

This model enables researchers to understand the nature of information retrieval. It also clarifies and integrates some of the key concepts related to information retrieval research. The planned-situational model can offer suggestions for the design of interactive information retrieval systems to support multiple information-seeking strategies and shifts in these strategies. It can also help researchers determine the system role and user involvement in supporting shifts in information-seeking strategies.

1. INTRODUCTION

The emergence of the Internet allows millions of users to access various types and formats of digital information regardless of their physical locations. End users normally are in a typical "information rich" environment in which people are exposed to, and make use of a variety of information resources in support of their work and personal needs. Information seeking is never an easy task. Their information problems require different information-seeking strategies. The new digital environment not only forces people to apply more than one type of information-seeking strategy, but also requires people to change from one information-seeking strategy to another in the information-seeking process. However, current information retrieval systems, such as Web search engines (e.g., Google), online databases (e.g., EbscoHost), online public access catalogueues (OPACs), and digital libraries (e.g., American Memory from Library of Congress) are designed mainly to support query formulation and limited browsing.

The digital age brings changes to information retrieval systems, users, information, and the environments in which users interact with systems. That also poses challenges for users to effectively retrieve information to accomplish their tasks and goals. The challenges for users can be summarized from the system side, user side, and information side.

From the system side, traditional information retrieval is supported by the two core processes: representation and comparison. In digital environments, term match—rather than concept match or problem match—is still a critical issue even though the search mechanism has been enhanced. information retrieval systems in digital environments do provide a variety of browsing mechanisms for users to explore information, but the query box is still the main channel for users to express their information needs. Users are limited by the search box, and most of the searches contain only one or two terms (Jansen & Pooch 2001). While users engage in multiple information-seeking strategies in digital environments (Drabenstott 2003; Fidel et al. 1999; Marchionini 1995; Vakkari, Pennanen, & Serola 2003; Wang, Hawk, & Tenopir 2000) and shifts in information-seeking strategies (Olah 200; Robins 2000), online information retrieval systems still focus on supporting searching-related strategies while offering some help with limited browsing. Interactivity is a fundamental characteristic of searching in digital environments. Users are able to interact with online information retrieval systems as well as their collections via multiple avenues. The inherent interactive nature of Web-based information retrieval systems poses a challenge for users. While users praise the ease-of-use of interfaces of online information retrieval systems, they are also concerned with the lack of control in effectively interacting with these systems. The simplified design of Web search engines has been transferred to other types of information retrieval system design. Researchers have paid more attention to ease-of-use of interface design and far less to user control. The existing online information retrieval systems do not support both ease-of-use and user control (Xie & Cool 2000, Xie 2003). Accordingly, the design of online information retrieval systems needs to consider user involvement and system role to facilitate user-system interaction (Bates 1990; White, Jose & Ruthven 2006; Xie 2003)

From the user side, any human being is potentially an end user in digital environments. For any given information retrieval system in the digital environment, universal access is an objective. Users could represent diverse user groups with diverse backgrounds. They could be heterogeneous in terms of their subject knowledge, system knowledge, information-seeking skills, and search styles. Users could also have different types of tasks. Even though they might look for the same information, different work or personal related tasks might lead them to retrieve information that requires different results to accomplish their tasks. Users might also exhibit different types of search strategies and behaviors. The question is how to support end users of online information retrieval systems who have different familiarity with the system environment, different information-seeking skills, different domain knowledge, different search tasks and goals, and different information-seeking strategies. In digital environments, users are able to access OPACs, Web search engines, online databases, and digital libraries for different types of information. Their past experience and background affect the way they interact with different types of information retrieval systems. They bring their individual mental models and search strategies for one type of information retrieval system to another one.

From the information side, users interact with information contained in information retrieval systems, in particular, retrieved results to find information to solve their problems; these results lead them to search for needed information or to find new ideas to reformulate their queries if the results fail to provide relevant information. Moreover, the Web offers a different searching environment for users; it contains a variety of information in content, format, and organization (Fidel et al. 1999; Jansen, Spink, & Saracevic 2000; Wang, Hawk, & Tenopir 2000). When users interact with the retrieved results, they not only have to make relevance judgments but also have to make authority and quality judgments (Rieh 2002; Xie 2006). The majority of existing information retrieval systems do not support users' evaluation of results, which is one of the key information-seeking strategies that users apply in their information seeking process. Nor are users willing to devote much time to evaluate results.

The changes on information retrieval systems, users and information contained in these information retrieval systems raise issues for researchers to understand users' information-seeking strategies in new digital environments and pose challenges for designers to design information retrieval systems that facilitate effective user-system interactions. A model that is able to show users' selection information-seeking strategies, in particular, shifts in information-seeking strategies is needed.

2. PLANNED-SITUATIONAL MODEL: SHIFTS IN INFORMATION-SEEKING STRATEGIES AND AFFECTING FACTORS

2.1 Major components of planned-situational model

The planned-situational model of interactive information retrieval (Figure 1) is developed to illustrate how plans and situations co-determine the selection of information-seeking strategies in the information seeking process. This model is an extension of Xie's (2000, 2002) previous model derived from an empirical study. The original model is further enhanced based on the existing general interactive information retrieval models (Belkin's episode model of interaction with text, 1996; Ingwersen's, cognitive model, 1992, 1996; Ingwersen and Järvelin's integrated IS& R framework, 2005; Saracevic's stratified model, 1997), influential user-oriented approaches in information seeking and retrieval (Talyor's levels of information need approach, 1968; Belkin's anomalous state of knowledge (ASK) hypothesis, 1980; Dervin's sense-making approach, 1983; Kuhlthau's information search process, 1991; Wilson's information-seeking context approach, 1999; and cognitive work analysis illustrated by Fidel & Pejtersen, 2004), specific models related to interactive information retrieval (Ellis' model of information-seeking behavior, Ellis, 1989, Ellis & Haugan, 1997; Bates' berrypicking approach, 1989; Vakkari's theory of the task-based information retrieval process, 2001; Pharo's model of information behaviour based on the search situation transition schema, 2004; Spink's extended interactive model by incorporating five types of interactive feedback, 1997; Hert's information retrieval interaction relation to the information-seeking process, 1997; and Wang, Hawk, and Tenopir's multidimensional model of user-Web interaction, 2000), plus empirical interactive information retrieval studies in different types of digital environments including OPACs, online databases, Web search engines, and digital libraries.

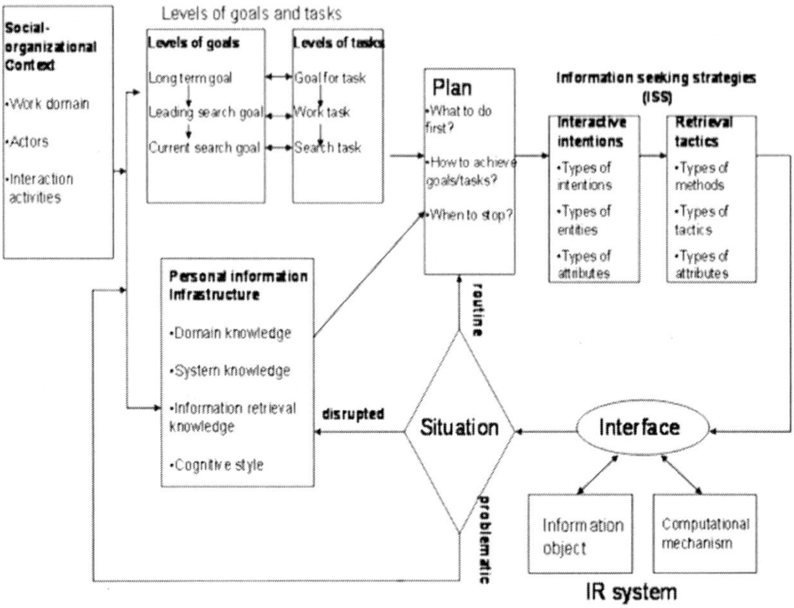

Figure 1. Planned-situational interactive information retrieval model

User goal and task is the driving force for people to look for information. In order to understand and support people's shifts in information-seeking strategies, we first need to define user goal and tasks as well as their relationships. Adapted from Daniels' (1986) classification of goals, Xie (2000, 2002) reconstructed user goals into four levels of hierarchical structure: long-term goal, leading search goal, current search goal, and interactive intention. In this structure, a leading search goal (e.g., writing a thesis) could be a subtask for someone to achieve his/her long term goal (e.g., achieve a master's degree); at the same time, it also could be a task for someone to work on his/her current search goal (e.g., find relevant literature about information-seeking strategies). In order to avoid the confusion between tasks and search tasks, the concept of work task is borrowed to represent a task that leads to the information searching (Borlund & Ingwersen 1997; Bystrom & Järvelin 1995; Ingwersen 1996; Ingwersen & Järvelin 2005; Vakkari 2001). Work task analysis, which focuses on specific tasks that people have to accomplish, what the constraints are, and what types of information sources are needed, is also a major component of cognitive work analysis (Pejtersen & Fidel 1998; Fidel & Pejtersen 2004).

Considering the situation of information seeking and retrieval, leading searching goals are comparable as work tasks while current search goals can be regarded as search tasks that are subtasks in the task performance. Correspondingly, interactive intentions which are subgoals that a user has to achieve in the process of accomplishing his or her current search goal are another level of subtasks. Users' search tasks/current search goals are influenced by work tasks/leading search goals. Simultaneously, long-term goals affect the work tasks/leading search goals in terms of what work tasks users might take on and how they might accomplish these work tasks. At the same time, search tasks/current search goals are influenced by work tasks/leading search goals, and they control both the kind of search results a user tries to obtain, and the means to obtain them.

At the same time, information seeking requires users to apply their knowledge and skills—what might be called "personal information infrastructures," such as their general cognitive abilities, their knowledge skills in relation to the problem/task domain, their knowledge and skills in general, their knowledge and skills specific to a system, and their knowledge and skills regarding information seeking (Marchionini 1995). In order to effectively interact with information retrieval systems, users need to have knowledge about the task that drives them to interact with information retrieval systems, knowledge of the information retrieval system that users interact with, and knowledge about how to interact. In addition, the cognitive styles of users determine how they might interact with information retrieval systems.

The social-organizational context defines the environment in which users interact with information retrieval systems. In the planned-situational model, context is mostly defined by the work domain that users interact with as suggested by cognitive work analysis (Rasmussen et al. 1994; Vicente 1999). The work domain, actors, and interaction activities and their properties are the essential parts of cognitive work analysis (Fidel & Pejtersen 2004). In interactive IR, users and information retrieval systems are the two partners. Users interact with information retrieval systems via interfaces of these systems. By interacting with the interfaces of information retrieval systems, users interact with the information objects stored in these systems as well as with the computational mechanisms of the software and hardware of the systems. Users' information-seeking strategies are also affected by the design of interfaces of information retrieval systems as well as the information objects stored in these systems in the following ways: 1) the design of overall user interface could also lead to more applications of certain strategies and less of others; 2) the availability or unavailability of certain features suggests whether users could engage in certain strategies; 3) the information objects stored in information retrieval systems that users interact with in general might determine the outcome of the current strategy, which, in turn, might affect their choices of next strategy.

Researchers have examined information-seeking strategies from different levels: 1) tactics/moves, such as Bates' (1979a, 1979b) information tactics, Shute & Smith's (1993) knowledge-based search tactics, Fidel's (1991) operational and conceptual moves, Shiri & Revie's (2003) cognitive and physical moves, Wildemuth's (2004) domain-knowledge-related search tactics, etc. 2) information-seeking strategies, such as concept-oriented strategies identified by Markey and Atherton (1978), Chen and Dhar's (1991) system-oriented strategies, browsing strategies versus analytical strategies discussed by Marchionini (1995), the plan strategies versus reactive strategies identified by Soloman (1997), Drabenstott's (2003) nondomain exerts' strategies, etc. 3) patterns, such as Ellis and Haugan's (1997) information seeking patterns of engineers and research scientists, etc. However, the identification of information-seeking strategies is still focused on the query formulation and reformulation. Identification of multiple information-seeking strategies that users engage in the information seeking process is essential. More research is needed to understand the nature of information retrieval in terms of not only how users act but also under what circumstances users shift their strategies in general information retrieval environment.

Information-seeking strategies comprise interactive intentions and retrieval tactics. Interactive intentions refer to subgoals that a user has to achieve in the process of accomplishing his or her current search goal/search task. Even though users have different leading search goals and current search goals, they share similar interactive intentions in the information retrieval process. Belkin et al. (1990) developed a set of intentions for people to find information in a variety of library environments. Xie (2000, 2002) expanded the intentions and named them "interactive intentions." Types of interactive intentions are further enhanced by incorporating empirical studies in digital environments such as OPAC, online databases, Web search engines, and digital libraries. Twelve types of interactive intentions are identified.

The modifications of types of interactive intentions with Xie's (2000, 2002) previous works can be summarized into the following aspects: 1) Modifications of interactive intentions are made considering interactive intentions derived from users' interaction with different types of online information retrieval systems. The following interactive intentions are added to the list: exploring, creating, modifying, monitoring, and disseminating in addition to the original ones, such as identifying, learning, keeping records, accessing, organizing, evaluating, and obtaining. Simultaneously, locating is deleted because it is more for locating items in libraries. Users are able to access and locate items in electronic systems altogether. "Finding" is replaced by "creating search statements" and "modifying search statements," since finding is a high level goal—a current search goal instead of an interactive intention. 2) Interactive intentions are further defined by entities and associated attributes. In this model, the author introduces nine new types of entities. In digital environments, information resource is no longer the only entity available. In order to effectively retrieve information, users have to interact with data/information, knowledge, concept/term, format, item/objects/site, process/status, location, system, and human. As part of interactive intentions, while entities refer to what users intend to acquire or work on, attributes specify the traits/elements of these entities. For example, the attributes for data/information are specific, common, general, and undefined, and the attributes for knowledge consist of domain, system and information retrieval knowledge.

Retrieval tactics are represented by methods and entities with attributes. Methods refer to the techniques users apply to interact with data/information, knowledge, concept/term, format, item/objects/site, process/status, location, system and humans. The types of methods include scanning, specifying, manipulating, tracking, surveying, selecting, comparing, extracting, acquiring, consulting, and trial-and-error. In Xie's previous research (2000, 2002), searching represents all the activities related to the identification of related items from collections of information retrieval systems. However, it is too general; moreover, the pre- and post- searching of management of concepts and process is missing from searching. Therefore, manipulating is added to represent query formulation, reformulation, and system customization, and surveying is introduced to review search process or examine current status. In addition, extracting is added to characterize tactics related to take out key information. The same types of entities and attributes for interactive intentions are also the dimensions for retrieval tactics. The integration and combination of these dimensions characterizes the retrieval tactics employed by users. Information-seeking strategies are constituted by interactive intentions and retrieval tactics. Here is an example of one type of information seeking strategy: learning domain knowledge by selecting descriptors of retrieved results. The first part is an interactive intention, and second part is a retrieval tactic. Multiple types of retrieval tactics can be applied in order to accomplish each type of interactive intention.

2.2 Shifts in information-seeking strategies determined by planned and situational aspects

Derived from cognitive point of view (Newell & Simon 1972), information-seeking behaviors are considered interrelated actions, and they are all part of a goal-related plan. A plan is a user's arrangement to complete a task, and it is not limited by the fixed actions. Applied Vera and Simon's (1993) perspective regarding how plans affect human actions, then plans influence information-seeking behavior in three ways: first, plans may be used to establish a set of intentions or subgoals along the route to accomplish their current search goals; second, plans may be used to determine corresponding information-seeking strategies that lead toward desired goals/tasks; third, plans also help in monitoring the search process to adjust the original plan. In that sense, three major elements can be characterized for a plan. First, what to do first? It refers to users' first interactive intention and its associated retrieved tactics. Second, how to achieve the goal/task? It includes a set of subgoals or interactive intentions that users have to accomplish in order to achieve their current search goals or search tasks. It consists of a set of plans that correspond to different types of situations or interactive intentions. More important, the plan also contains arrangements about how to monitor the retrieval process. Third, when to stop? It is related to users' decisions about when they quit their retrieval process after obtaining complete information, enough information, partial information, or just by frustration.

A plan is co-determined by levels of user goals/tasks, personal information infrastructure, and the social-organizational context. As discussed above, levels of user goals/task determine user plans. The dimensions of leading search goals/work tasks define the work tasks in terms of stages of the task, timeframe of the task, and the nature of the task (the structuredness of task, users' familiarity with the task, and situations of the tasks). The user goals/tasks structure also depicts dimensions of search tasks: origination of tasks, types of tasks, and domain of tasks. These dimensions assist users consciously and unconsciously in selecting appropriate plans for their tasks and, at same time, monitor the information retrieval process. Users' domain knowledge, system knowledge, information retrieval knowledge, and cognitive styles assist users in determining their actions and plans under different circumstances. The social-organizational context in which the interaction takes places defines the priority and limitations of what users can do in their selections of plans and selections of information-seeking strategies in the retrieval process. Xie (2006) identified and validated some of these dimensions of plans by investigating how people seek information in a corporate setting.

Contrasted to cognitivist view, theory of situated action (Suchman, 1987), drawn from recent developments in the social sciences, argues that the coherence of action is not adequately explained by either preconceived cognitive schema or institutionalized social norms. The organization of situated action is an emergent property of moment-by-moment interactions between actors and the environment that they are interacting with. A situation is a user's perception of a specific moment based on his/her evaluation of the interaction with an information retrieval system and his/her plan. The elements of situations were well defined by Hert (1997): elements associated with the problem (subject area, specific requirement, point in the process, group projects or not), elements associated with system response (nature of retrieved sets, status messages, features of individual entries), and elements associated with users (knowledge, attitudes, expectations, emotions, etc.).

A situation is the product of interactions between users and information retrieval systems. Situations arise within each social-organizational context (Sonnenwald, 1999). Under the social and organizational context and driven by their levels of goals and tasks, users apply their own information infrastructure to interact with information retrieval systems via interfaces of these systems. Users' interactions with information retrieval systems, especially the objects presented by the systems, create different situations in the interactive information retrieval process. In addition, users' plans are part of the situations since users' perception of a specific time-space is the result of the assessment of their plans based on the outcomes of interactions with information retrieval systems.

To some extent, planned and situational aspects are intertwined with each other. For example, information seeking requires users to apply their personal information infrastructures, such as their general cognitive abilities, knowledge skills in relation to the problem/task domain, knowledge and skills specific to a system, and knowledge and skills regarding information seeking. Users' knowledge also includes a set of plans for different situations with a set of interactive intentions and associated retrieval tactics. More important, these plans represent users' conscious and unconscious actions toward familiar and unfamiliar situations. Users normally employ their personal information infrastructure to: 1) represent their problems/tasks; 2) establish a set of subgoals/subtasks to fulfill the overall goals/tasks; and 3) develop techniques and strategies to seek required information. At the same time, users' personal information infrastructures are also developed during the information-seeking process when users gain knowledge and skills in order to adapt to different situations and solve problems. Derived from users' knowledge and skills, their information-seeking strategies are the stratification of their experiences according to relevance and typicality. Therefore, information-seeking strategies are the products of plans and situations. The information retrieval process is determined by both planned and situational aspects.

Different types of situations lead to different types of shifts in information-seeking strategies. Two types of situations identified by Schutz and Luckmann (1973): routine situation and problematic situation. These two situations represent the most frequently occurred situations that users encounter in their information seeking process. In a routine situation, the situation can be determined sufficiently with the aid of habitual knowledge. After assessing the plan and the current situation, as planned, a user normally moves to the next interactive intention, and selects its corresponding retrieval tactics to fulfill his/her new interactive intention. In a problematic situation, after associating the elements of situations with his/her knowledge and plans, further clarification of the open elements of the situation is required. The situational aspects have impact on which facets of knowledge and skills are brought to bear, modifications and re-arrangements of the set of interactive intentions and retrieval tactics, and a clearer understanding and adjustment of levels of user goals. Accordingly, an adjusted plan is formed to shift information-seeking strategies.

A third type of situation, which can be named "disruptive situation," emerges in the context of interactive IR. In disruptive situations, users do not encounter any problems, but they are distracted in the process of fulfilling their original current search goals/search tasks. For example, they might see something that holds their interest that is not relevant to their current search goals/tasks. They explore the new interest for a while, and then come back to their original current search goals/tasks. In the level of interactive intention shifts, users have to apply their personal information infrastructure to go back their plan for the next appropriate interactive intentions and its corresponding retrieval tactics in order to accomplish their original current search goals/search tasks. To sum up, shifts in information-seeking strategies are co-determined by planned and situational aspects.

3. CONCLUSION: IMPLICATIONS AND FURTHER RESEARCH

The main theoretical contribution of the model is that it enables researchers to understand the nature of IR. It is a challenge for people to retrieve useful information to accomplish their work or personal related tasks and goals in the digital age. The planned-situational model presents how plans and situations co-determine users' selection of information-seeking strategies during their interactions with information retrieval systems. Levels of user goals/tasks and their dimensions affect the way that users interact with information retrieval systems and the scope of the information they intend to retrieve. The identification of types of interactive intentions with corresponding retrieved tactics represents multiple information-seeking strategies users have to apply in order to achieve their goals/tasks. Multiple information-seeking strategies are products of plans and situations. This model clarifies and integrates some of the key concepts related to information retrieval research, such as tasks, goals, problematic situations, search tactics, information-seeking strategies, etc.

Practically, the planned-situational model can offer suggestions for how to design interactive information retrieval systems to support multiple information-seeking strategies and shifts in information-seeking strategies. Moreover, the model can help researchers to answer the essential question related to the system design: "What are the system role and user involvement in supporting and applying multiple information-seeking strategies across different types of information retrieval systems? Bates (1990) emphasized the need to delegate clear responsibilities for system and user involvements in the information retrieval process. The success of user-system interaction depends on the collaboration of both partners. Brajnik, et al. (2002) argued that neither user nor the system could solve information problems individually, and they proposed a collaborative coaching approach where users are in charge of search sessions while systems offer suggestions. White, Jose and Ruthven (2006) pointed out that it is important to understand what users would like to control and what they would be willing to allow systems to control. Xie (2003) concluded in an empirical study that user involvement focuses on how to make conceptual judgments and decisions while the system role concentrates on how to enhance user's knowledge structure and facilitate them to make a variety of judgments and decisions. Users engage in multiple information-seeking strategies in the information retrieval process, but not all the strategies require the same level and same type of involvement from users and systems. While users take leading roles in identifying, exploring, learning, creating, modifying, and evaluating strategies, systems mainly deal with the monitoring, accessing, organizing, keeping records, obtaining, and disseminating strategies.

This model needs to be further validated and enhanced by empirical studies. These studies ought to investigate real users with real problems in real settings. In particular, different types of users with different types of work or personal related tasks and goals searching for different types of information retrieval systems or information resources in their natural settings are required for the studies.

ACKNOWLEDGEMENTS

The author would like to thank the anonymous reviewers' insightful comments.

REFERENCES

1. Bates, M. J. (1979a). Information search tactics. *Journals of the American Society for Information Science,* **30**(4), 205-214.

2. Bates, M. J. (1979b). Idea tactics. *Journals of the American Society for Information Science,* **30**(5), 280-289.

3. Bates, M. J. (1989). The design of browsing and berrypicking techniques for the online search interface. *Online Review,* **13**, 407-424.

4. Bates, M. J. (1990). Where should the person stop and the information search interface start? *Information Process & Management,* **26**(5), 575-591.

5. Belkin, N. J. (1980). Anomalous states of knowledge as a basis for information retrieval. *Canadian Journal of Information Science*, **5**, 133-143.

6. Belkin, N. J. (1996). Intelligent information retrieval: Whose intelligence? *Proceedings of the Fifth International Symposium for Information Science* (pp. 25–31). Konstanz: Universita¨tsverlag Konstanz.

7. Belkin, N. J., Chang, S., Downs, T., Saracevic, T., & Zhao, S. (1990). Taking account of user tasks, goals and behavior for the design of online public access catalogueues. In D. Henderson (Ed.), *Proceedings of the 53rd ASIS Annual Meeting* (pp. 69-79). Medford, NJ: Learned Information Inc.

8. Borlund, P., & Ingwersen, P. (1997). The development of a method for the evaluation of interactive information retrieval systems. *Journal of Documentation*, **53**(3), 225-250.

9. Brajnik, G., Mizzaro, S., Tasso, C., & Venuti, F. (2002). Strategic help in user interfaces for information retrieval. *Journal of the American Society for Information Science and Technology*, **53**(5), 343-358.

10. Bystrom, K., & Järvelin, K. (1995). Task complexity affects information-seeking and use. *Information Processing & Management*, **31**(2), 191-213.

11. Chen, H., & Dhar, V. (1991). Cognitive processes as a basis for intelligent retrieval system design. *Information Processing & Management*, **27**, 405-432.

12. Daniels, P. (1986). *Developing the user modeling function of an intelligent interface for document retrieval systems.* Unpublished doctoral dissertation. Department of Information Science, the City University, London, England.

13. Dervin, B. (1983, May). *An overview of sense-making research: Concepts, methods and results to date.* Paper presented at the

International Communication Association Annual Meeting, Dallas, TX.

14. Drabenstott, K. M (2003) Do nondomain experts enlist the strategies of domain experts? *Journal of the American Society for Information Science and Technology*, **54**, 836-854.

15. Ellis, D. (1989). A behavioural approach to information retrieval system design. *Journal of Documentation*, **45**(3), 171-212.

16. Ellis, D., & Haugan, M. (1997). Modelling the information seeking patterns of engineers and research scientists in an industrial environment.*Journal of Documentation*, **53**(4), 384-403.

17. Fidel, R. (1991). Searchers' selection of search keys: I. The selection routine. II. Controlled vocabulary or free-text searching. III. Searching styles.*Journal of the American Society for Information Science*, **42**, 490-527.

18. Fidel, R., Davies, R. K., Douglass, M. H., Holder, J. K., Hopkins, C. J., Kushner, E. J., et al., (1999). A visit to the information mall: Web searching behavior of high school students. *Journal of the American Society for Information Science*, **50**(1), 24–37.

19. Fidel, R., & Pejtersen, A. M. (2004). From information behavior research to the design of information systems: The cognitive work analysis framework. *Information Research*, **10**(1). Retrieved February 28, 2006, from http://informationr.net/ir/10-1/paper 210.html

20. Hert, C. A. (1997). *Understanding information retrieval interaction: Theoretical and practical implications*. Greenwich, CO: Ablex Pub.

21. Ingwersen, P. (1992). *Information retrieval interaction*. London: Taylor Graham.

22. Ingwersen, P. (1996). Cognitive perspectives of information retrieval interaction: Elements of a cognitive information retrieval theory. *Journal of Documentation*, **52**(1), 3-50.

23. Ingwersen, P., & Järvelin, K. (2005). *The turn: Integration of information.seeking and retrieval in context.* Heidelberg: Springer.

24. Jansen, B. J., & Pooch, U. (2001). A review of Web searching studies and a framework for future research. *Journal of the American Society for Information Science and Technology,* **52**(3), 235-246.

25. Jansen, B. J., Spink, A., & Saracevic, T. (2000). Real life, real users, and real needs: A study and analysis of user queries on the web. *Information Process & Management,* **36**, 207–227.

26. Kuhlthau, C. C. (1991). Inside the search process: Information seeking from the user's perspective. *Journal of the American Society for Information Science,* **42**, 361-371.

27. Marchionini, G. (1995). *Information seeking in electronic environments.* Cambridge: Cambridge University Press.

28. Markey, K, & Atherton, P. (1978). *ONTAP: On-line training and practice manual for ERIC database searchers.* ERIC Clearinghouse on Information Resources, Syracuse University.

29. Newell, A., & Simon, A. (1972). *Human problem solving.* Englewood Cliffs, NJ: Prentice-Hall.

30. Olah, J. (2005). Shifts between search stages during task-performance in mediated information seeking interaction. *Proceedings of the 68th Annual meeting of the American Society for Information Science and Technology,* electronic.

31. Pejtersen, A. M., & Fidel, R. (1998). *A framework for work-centred evaluation and design: A case study of information retrieval on the Web.*Paper presented at the MIRA workshop, Grenoble, March 1988. Retrieved December 20, 2006, from http://www.dcs.gla.ac.uk/mira/workshops/grenoble/fp.pdf

32. Pharo, N. (2004). A new model of information behaviour based on the search situation transition schema. *Information Research,* **10**(1) paper 203. Retrieved November 15, 2006 from http://InformationR.net/ir/10-1/paper203.html

33. Rasmussen, J., Pejtersen, A. M., & Goodstein, L. P. (1994). *Cognitive systems engineering*. New York, NY: John Wiley & Sons, Inc.

34. Rieh, S. Y. (2002). Judgment of information quality and cognitive authority in the Web. *Journal of the American Society for Information Science and Technology*, **53**(2), 145-161.

35. Robins, D. (2000). Shifts of focus on various aspects of user information problems during interactive information retrieval. *Journal of the American Society for Information Science*, **51**(10), 913-928.

36. Saracevic, T. (1997). The stratified model of information retrieval interaction: extension and applications. *Proceedings of the 60th Annual meeting of the American Society for Information Science* (pp. 313-327). Medford, NJ: Learned Information Inc.

37. Schutz, A., & Luckmann, T. (1973). *The Structure of the life world.* Evanston, IL: Northwestern University Press.

38. Shiri, A. A., & Revie, C. (2003). The effects of topic complexity and familiarity on cognitive and physical moves in a thesaurus-enhanced search environment. *Journal of Information Science*, **29**, 517-526.

39. Shute, S. J., & Smith, P.J. (1993). Knowledge-based search tactics. *Information Processing & Management*, **29**, 29-45.

40. Soloman, P. (1997). Children's information retrieval behavior: A case analysis of an OPAC. *Journal of the American Society for Information Science*, **44**(5), 245-264.

41. Sonnenwald, D.H. (1999). Perspectives of human information behaviour: contexts, situations, social networks and information horizons. In*Exploring the Contexts of Information Behaviour: Proceedings of the second international conference on research in information needs, seeking and use in different contexts* (Sheffield, UK, 13-15 August 1998). Taylor Graham, 176-190.

42. Spink, A. (1997). Study of interactive feedback during mediated information retrieval. *Journal of the American Society for Information Science,***48**(5), 382-394.

43. Suchman, L.A. (1987). *Plans and situated actions: The problems of human-machine communication.* Cambridge: Cambridge University Press.

44. Taylor, R. (1968). Question-negotiation and information seeking. *College and Research Libraries,* **29**, 178-194.

45. Vakkari, P. (2001). A theory of the task-based information retrieval process.*Journal of Documentation,* **57**, 44-60.

46. Vakkari, P., Pennanen, M., & Serola, S. (2003). Changes in search terms and tactics while writing a research proposal: A longitudinal case study.*Information Processing & Management,* **39**, 445–463.

47. Vera, A. H., & Simon, H. A. (1993). Situated action - A symbolic interpretation. *Cognitive Science,* **17,** 7-48.

48. Vicente, K. J. (1999). *Cognitive work analysis: Toward safe, productive, and healthy computer-based work.* Mahwah, NJ: Lawrence Erbaum Associates.

49. Wilson, T. D. (1999). Models in information behavior research. *Journal of Documentation,* **55**, 249-270.

50. Wang, P., Hawk, W.B., & Tenopir, C. (2000). Users' interaction with World Wide Web resources: An exploratory study using a holistic approach.*Information Processing & Management,* **36**, 229–251.

51. Wildemuth, B.M. (2004). The effects of domain knowledge on search tactic formulation. *Journal of the American Society for Information Science & Technology,* **55**(3), 246-258.

52. White, R. W., Jose, J. M., & Ruthven, I. (2006). An implicit feedback approach for interactive information retrieval. *Information Processing & Management,* **42**(1), 166-190.

53. Xie, H. (2000). Shifts of interactive intentions and information-seeking strategies in interactive information retrieval. *Journal of the American Society for Information Science*, **51**(9), 841-857.

54. Xie, H. (2002). Patterns between interactive intentions and information-seeking strategies. *Information Processing & Management*, **38**(1), 55-77.

55. Xie, H. (2003). Supporting ease-of-use and user control: Desired features and structure of Web-based online information retrieval systems.*Information Processing & Management*, **39**(6), 899-922.

56. Xie, H. (2006). Understanding human-work domain interaction: Implications for the design of a corporate digital library. *Journal of the American Society for Information Science and Technology*, **57**(1), 128-143.

57. Xie, H., & Cool, C. (2000). Ease-of-use versus user control: An evaluation of Web and non-Web interfaces of online databases. *Online Information Review*, **24**(2): 102-115.

8

TEXTUAL AND CHEMICAL INFORMATION PROCESSING: DIFFERENT DOMAINS BUT SIMILAR ALGORITHMS

Peter Willett

Department of Information Studies and Krebs Institute for Biomolecular Research, University of Sheffield, Sheffield S10 2TN, UK

ABSTRACT

This paper discusses the extent to which algorithms developed for the processing of textual databases are also applicable to the processing of chemical structure databases, and vice versa. Applications discussed include: an algorithm for distribution sorting that has been applied to the design of screening systems for rapid chemical substructure searching; the use of measures of inter-molecular structural similarity for the analysis of hypertext graphs; a genetic algorithm for calculating term weights for relevance feedback searching for determining whether a molecule is likely to exhibit biological activity; and the use of data fusion to combine the results of different chemical similarity searches.

1. INTRODUCTION

Over the years, researchers in information retrieval (IR) have developed many different techniques for the processing of databases of textual information. Established examples of such techniques include document clustering, relevance feedback, stemming algorithms and text compression, and there are many other emerging applications, such as categorisation, event detection and information filtering (Spark Jones and Willett, 1997). One such application is multimedia retrieval, where there is much current interest in extending algorithms and data structures developed for processing textual databases, for the storage and retrieval of speech, image and video data (Maybury, 1997). This paper argues that at least some of the algorithms that are used in textual information retrieval can also be applied to another type of data, *viz* the two-dimensional (2D) and three-dimensional (3D) chemical structure data that forms one of the principal components of chemical information systems (Ash *et al.*, 1991). These systems were first developed principally for archival purposes, but now play an important role in research programmes to discover novel bioactive molecules for the pharmaceutical and agrochemical industries (Martin and Willett, 1998).

The University of Sheffield has been involved in research on both chemical and textual information retrieval for many years (Lynch and Willett, 1986). These studies have led us to believe that both areas have much to offer, with research in one providing a fertile source of ideas for research in the other. In some cases, the relationship is obvious, with algorithms and data structures being transferable with little or no change from one application to another; while in other cases, the relationship is less direct, involving more a general view of the sorts of information processing techniques that are required, rather than a direct transfer of technology. Here, we consider the former, more direct, type of relationship.

There are clear similarities in the ways that chemical and textual database records are characterised. The documents in a text database are each typically indexed by some small number of keywords, in just the same way as the 2D or 3D molecular representations in a chemical database are each characterised by some small number of substructural features chosen from a much larger number of potential attributes (as discussed further in the next section of this paper). Moreover, both types of attribute follow a well-marked Zipfian distribution, with the skewed distributions that characterise the frequencies of occurrence of characters, character substrings and words in text databases being mirrored by the comparable distributions for the frequencies of chemical moieties. Thus, the overwhelming majority of all of the many millions of molecules that have ever been made contain the element carbon but even the tenth most frequent element, iodine, occurs only about one thousandth as frequently, with the great majority of the elements having vanishingly small frequencies of occurrence; similar distributions are observed for other types of chemical substructure (Lynch, 1977). These shared characteristics mean that the two types of database are amenable to efficient processing using the same type of file structure. Thus one of the first examples of what would now be referred to as text signature searching (Barton*et al.*, 1974) arose from previous studies of chemical bit-string processing (Adamson *et al.*, 1973), and similar comments apply to the use of an inverted file for rapid database clustering (Willett, 1981, 1982). Finally, in just the same way as a document either is, or is not, relevant to some particular user query, so a molecule is active, or is not active, in some particular biological test, thus allowing comparable performance measures to be used to assess search effectiveness in the two types of retrieval system (Edgar *et al.*, 1999).

These similarities mean that it is often possible to apply similar algorithms to the two different sorts of database, as we describe in some detail below. That said, there are obvious differences, most obviously in the semantics of the representations that are used. A 2D chemical structure diagram bears a much closer relationship to the molecule it describes than does the set of words comprising a textual document, and this relationship is still stronger when, as is increasingly the case, a 3D structure with XYZ atomic co-ordinate data is available for a molecule (Martin and Willett, 1998). Both the structure diagram and the co-ordinates can be regarded as direct manifestations of the underlying wave equations that describe a molecule, and it has thus proved possible to develop powerful simulation techniques to predict the activities and properties of molecules from a knowledge of their 2D or 3D structure. Many of these molecular modelling tools have no direct textual equivalent, as the use of natural language raises a host of linguistic problems that do not arise in the chemical context (although it should be noted that linguistic parsing and recognition techniques can be used to represent and search the generic chemical structures that occur in chemical patents (Barnard *et al.*, 1984; Welford *et al.*, 1981)).

The remainder of this paper discusses several applications to support the belief that there may be at least some degree of overlap between the techniques used to process chemical and textual databases. The first application, which is taken from a previous paper discussing a potential relationship between these two domains (Willett, 1997), is that of chemical substructure searching; this section also introduces the basic components of chemical information systems, thus providing some of the necessary background for the more recent applications discussed in the following sections.

2. DESIGN OF SCREENING SYSTEMS FOR CHEMICAL SUBSTRUCTURE SEARCHING

Chemical information systems have historically represented molecules by means of their 2D chemical structure diagrams: these are encoded as labelled graphs in which the nodes and edges of a graph encode the atoms and bonds of a molecule (Ash *et al.*, 1991). Searches for substances that contain a specific partial structure, such as a cephalosporin ring, can then be effected by graph-matching techniques that permit the identification of all occurrences of the user. s query substructure in each database structure (Barnard, 1993). The time-consuming nature of these *subgraph isomorphism* searches means that an initial *screening*stage is required that can eliminate the great bulk of the database from subsequent processing (in just the same way as a text signature search is used to reduce the amount of pattern matching that is required in serial searches of text databases). The question then arises as to what sorts of substructural characteristics should be used as indexing keys in the screening search.

The Zipfian distribution of the occurrences of the elements has been mentioned previously, this implying a vast divergence in the discriminatory powers of searches of chemical databases that are based on elemental type. Work in Sheffield by Lynch and his co-workers in the early Seventies (see, *e.g.*, Adamson *et al.*, 1973) showed that improved discriminatory power could be obtained by the use of more sophisticated indexing keys that were based on variably-sized chemical *fragment substructures, i.e.*, groups of atoms and bonds. The fragments were chosen so as to occur with approximately equal frequencies of occurrence in the database that was to be searched, an idea that now forms the basis for many current systems for 2D substructure searching (Barnard, 1993).

It was soon realised that the concept of equifrequency was of great generality, and this led to an extended investigation of the application of equifrequency ideas to the searching and processing of text databases (Lynch, 1977): here, we consider the studies that were carried out on the approach to sorting normally refererd to as *distribution sorting* (Cooper *et al.*, 1980; Cooper and Lynch, 1984). The suggested approach involves sorting a very large file in two stages: first, an initial, approximate sorting stage that sub-divides the file into an ordered series of sub-files; and then an exact sort of each of the sub-files. To those of us familiar with the old days when catalogue cards had to be sorted by hand, the initial stage is analogous to dividing the cards into the sub-files A to D, E to K, L to R and S to Z. In the present context, each of these rough pigeonholes represents a separate sub-file on disk and each of the ranges is chosen to ensure that approximately equal numbers of records are allocated to each such sub-file, with the aim of maximising the efficiency of the final, in-core sorts of these sub-files.

The 3D structure of a molecule plays a vital role in determining its biological activity. The "lock-and-key" theory suggests that a molecule may be able to act as a drug if it can fit into the active site of a protein in much the same way as a key fits into a lock (Martin and Willett, 1998), and there is thus great interest in being able to identify molecules that are appropriately "key-shaped". This is done by searching for *pharmacophores, i.e.*, the patterns of atoms in 3D space that are thought to be responsible for a molecule binding to an active site in a protein molecule. When our studies of pharmacophore searching started in the mid-Eighties, it was soon realised that the graph-theoretic methods developed for 2D substructure searching were also applicable here, with a 3D molecule being described by a graph in which the nodes were the atoms and the edges were the inter-atomic distances (Willett, 1995). However, the implementation of subgraph isomorphism matching on such 3D chemical graphs is still more demanding of computational resources than is 2D substructure searching, with a consequent need for the development of efficient methods of screening.

The graphs representing 2D chemical molecules are composed of atoms and bonds, and this is reflected in the compositions of the screens that are used for 2D substructure searching. It thus seemed appropriate to investigate screens for 3D searching based on the information contained in 3D chemical graphs, *i.e.*, atoms and inter-atomic distances, and we decided to evaluate screens consisting of a pair of atoms and an inter-atomic distance range. Thus, a screen might represent the presence within a molecule of an oxygen atom and a nitrogen atom separated by a distance range, *e.g.*, between 5 and 7 Ångstroms. Initial experiments demonstrated the vastly skewed distributions of distances that characterise databases of typical 3D molecules and there was thus a need to identify such inter-atomic screens so that they occurred with approximately equal frequencies of occurrence (Jakes and Willett, 1986).

The approach adopted involved varying the width of the distance range associated with each pair of atoms, so that highly populated parts of the distance distribution were associated with very narrow ranges, while less populated parts of this distribution, or very infrequently occurring pairs of atoms, were associated with more extended ranges. The algorithm that was finally developed was a simple modification of the distribution sorting algorithm described above with, in essence, the distance ranges here corresponding to the character ranges that underlie the text-sorting application (Cringean *et al.*, 1990). This proved to be both effective and efficient, and screens based on the distances between pairs of atoms now form the basis for nearly all existing systems for 3D substructure searching; indeed, the same basic methodology can be used to characterise the valence or dihedral angular relationships that exist between sets of three or four atoms, respectively, thus allowing database searches to be carried out that involve the specification of both distance and angular information. More recently, we have demonstrated that analogous procedures can be used to search databases where account is taken of the fact that most 3D molecules are not completely rigid, but are actually in a state of constant flux (so that the "keys" referred to previously might be considered as being more akin to a jelly than to a piece of rigid metal) (Willett, 1995). This further application demonstrates clearly the generality of equifrequency-based approaches for database processing.

3. MANIPULATION OF HYPERTEXT GRAPHS USING MEASURES OF INTER-MOLECULAR STRUCTURAL SIMILARITY.

Text retrieval systems were initially based upon the Boolean retrieval model, but the systems were soon extended to permit best match searching (in which the documents are ranked in decreasing order of similarity to the query) (Spark Jones and Willett, 1997). A similar progression has occurred in chemical information systems, with the substructure searching systems described above increasingly being complemented with facilities for what is referred to as*similarity searching*. This generally involves the specification of an entire query molecule, the *target structure*, rather than the partial structure that is required for substructure searching. The target is characterised by one or more structural descriptors that are compared with the corresponding sets of descriptors for each of the molecules in the database to find those *nearest neighbour* molecules that are most similar to the target structure. Two near-contemporaneous studies in the mid-Eighties demonstrated that counts of the numbers of fragment substructures common to a pair of molecules provided a computationally efficient, and surprisingly effective, basis for quantifying the degree of structural resemblance between the two molecules under consideration (Carhart *et al*, 1985; Willett *et al.*, 1986). Specifically, the use of a simple association coefficient (usually the Tanimoto coefficient) in conjunction with the lists of screens associated with the target structure and each of the database structures provided a simple way of investigating inter-molecular structural similarities. Such fragment-based methods for similarity searching are now widely used as a complement to the established routines for substructure searching (Willett *et al.*, 1998) and we have since demonstrated that these measures are also applicable to the comparison of textual, rather than chemical, graphs. Specifically, as described below, we have used these measures to determine the degree of consistency with which hypertext documents are created; other relationships between textual and chemical similarity measures are discussed by Willett (1997).

The creation of the intra-document links between the individual components of a hypertext document is a difficult, and time-consuming, task, but one in which human intervention has commonly been thought necessary if the semantic relationships that exist between the components of the document are to be made explicit. A similar view has prevailed for many years with regard to the indexing of documents in IR systems, where the existence of well-established systems for automatic indexing has not prevented the widespread use of trained library and information specialists for indexing and classifying documents prior to their incorporation in an online database. The importance of the indexing task in IR has led to many studies of *inter-indexer consistency, i.e.*, of the extent to which agreement exists among different indexers on the sets of index terms to be assigned to individual documents. These studies (as reviewed, *e.g.*, by Markey (1984)) have consistently concluded that recorded levels of consistency vary markedly, and that high levels of consistency are rarely achieved; similar examples of manual inconsistency are provided by related tasks, such as query formulation and the assessment of document relevance (Salton, 1989). The insertion of links in hypertext documents may be viewed as being analogous to the assignment of index terms to such documents, and we hence undertook a study to determine the extent to which different people produce similar link structures for the same hypertext documents (Ellis *et al.*, 1994, 1996).

The hypertext documents were generated from five printed full-text documents, each a thesis, journal article or book written by a member of the Department of Information Studies at the University of Sheffield. Five copies were made of each of the chosen documents, and each of the twenty-five resulting copies was allocated to a different student volunteer from the Department. The volunteers were instructed in the use of an interactive system that allowed them to create explicit representations of links between paragraphs whose contents they decided were related, and thus to create hypertext versions of the source documents. These hypertexts were stored as graphs in which the nodes represented portions of a text (specifically paragraphs in our work but section-based or sentence-based portions could also have been used), with an edge linking a pair of nodes if the human linker had created a link between the corresponding paragraphs. Pairs of these graphs (describing the same textual document but processed by different volunteers) were then compared using a range of similarity measures, most of which were based on those used for chemical similarity searching. For example, a commonly used type of chemical fragment is the *augmented atom*, which consists of an atom and the atoms that are bonded to it, and it is possible to generate a comparable hypertext fragment consisting of a paragraph and those paragraphs that are linked to it. A measure of the similarity between two hypertext graphs can then be calculated in terms of the number of such fragments that the hypertexts have in common, and the resulting similarity is taken as a measure of the degree of inter-linker consistency.

Five hypertext versions were produced for each source document, giving a total of ten possible pairs of sets of hypertext links for that document (*i.e.*, fifty possible pairs for the entire dataset). Similarity values were calculated using many different combinations of graph representation, fragment descriptor and similarity coefficient. When this was done, it was possible to draw a simple, unequivocal conclusion: that levels of inter-linker consistency are generally low, but can also be quite variable. It would hence seem that different people tend to have very different views of the semantic relationships that exist among the components of a full-text document, with the possibility that different people will tend to impose different hypertext link structures on the same source documents. This lack of agreement leads one to question the effectiveness of manually-assigned links in supporting the browsing strategies of subsequent hypertext readers, since if inter-linker consistency is found to be low, then linker/reader consistency (*i.e.*, the level of agreement as to the semantic relationships that exist between the components of the text, and that therefore should explicitly be represented by links that the reader then has the opportunity to follow) can also be expected to be low. Indeed, our results provide some justification for the numerous and continuing efforts to develop techniques for the automatic generation of hypertext links (see, *e.g.*, Crestani and Melucci, 1998) where the set of links created by a particular technique will be produced in a consistent manner, to which users might become accustomed as they browse and which might well be no worse than those created by human indxers.

4. A GENETIC ALGORITHM FOR CALCULATING RELEVANCE FEEDBACK AND SUBSTRUCTURAL ANALYSIS WEIGHTS.

The calculation of weights that describe the importance of document and query terms is an important component of best-match text IR systems. The most effective results are obtained from using relevance feedback information (Robertson and Sparck Jones, 1976; Salton and Buckley, 1990) but efforts continue to develop new weighting schemes that might further increase search effectiveness. This striving for improved performance has raised the question of what is the*upperbound, i.e.*, the maximum possible level, that can be achieved by a particular retrieval strategy. We have hence developed a genetic algorithm (hereafter a GA) to investigate the upperbound that can be achieved by ranked-output retrieval systems that employ weighted query terms. Indeed, the GA was developed not only for this purpose but also for calculating weights reflecting the extent to which particular types of molecular feature contribute to the biological activity of molecules (using the analogies between indexing terms and structural features, and between query relevance and biological activity that we have noted in the introduction to this paper). We will illustrate the operation of this GA by discussing the textual application (Robertson and Willett, 1996) before proceeding to an account of the results obtained in our chemical experiments (Gillet *et al.*, 1998).

A GA operates on *chromosomes,* which are linearly-encoded representations of possible solutions to the problem under investigation, with each element of a chromosome describing some particular component of the encoded solution. A *fitness function* is used to calculate a numeric score that measures how "good" a solution is represented by each chromosome. The set of chromosomes at any particular stage of the processing is referred to as the current *population,* and the chromosomes in this generation are processed by *genetic operators* to yield the population that corresponds to the subsequent generation. The two principal operators are *crossover,* which combines parts of the fittest chromosomes, and *mutation,* which introduces new information at random. The chromosomes in the new generation are evaluated by means of the fitness function and the genetic operators invoked again, this process continuing until convergence is achieved, *i.e*, until there is no increase in the average fitness of each successive generation.

Each chromosome in our text GA contains a set of elements, with one element for each of the terms comprising the query and with each such element containing a weight for one of these terms. The weights for those query terms that occur in a particular document are summed, the documents are ranked in decreasing order of these sums of weights, and a cut-off is then applied to the ranking to retrieve some fixed number of the top-ranked documents. Our experiments have used standard document test collections for which full relevance data is available, and it is hence possible to determine the recall of the search by noting the number of relevant documents that have been retrieved above the cut-off. These recall values are used as the fitness function for the GA. Once termination has occurred, *i.e.*, the recall values have ceased to increase, the term weights are noted for the chromosome that has the largest fitness. Thus, if the GA has successfully explored the full range of possible weights, then the final weights provide an upperbound to the retrieval performance obtainable for that query using relevance weighting of single terms.

The experiments involved the following seven document test collections: Keen (800 documents and 63 queries), Cranfield (1400 documents and 225 queries), Harding (2472 documents and 65 queries), Evans (2542 documents and 39 queries), LISA (6004 documents and 35 queries), SMART (12694 documents and 77 queries) and UKCIS (27361 documents and 182 queries). The GA was run for each query in each collection, and the final set of weights used to calculate recall values for the top-10, top-20 and top-50 documents. The F4 retrospective relevance weights of Robertson and Spark Jones (1976) were used for comparison with the GA weights since the former have been found to provide a consistently high level of performance in previous studies of relevance feedback searching.

Table 1 (below) summarises the results obtained when the top-20 documents were retrieved from a ranking; many other results are presented and discussed by Robertson and Willett (1996). An inspection of this table shows that (as would be expected) the GA weights generally perform better than the F4 weights, in terms of both average recall and numbers of queries where one was superior to the other. However, there are generally many searches where the two approaches give the same level of performance (to two decimal places in mean recall), and often at least a few queries (and a fair number in the case of the Harding collection) where the F4 search is better. Where there are differences, in either direction, between the two weights, these are overwhelmingly due to one of the weights finding a single additional relevant document above the cut-off position in the ranking: thus, the two approaches give very similar levels of performance. A study of those few queries where the F4 weights were (unexpectedly) superior suggested that these queries were ones in which at least some of the F4 weights were negative (this corresponding to terms that occur frequently in a collection but only infrequently in the relevant documents). The initial version of the GA did not allow for the possibility of negative weights: when it was modified to allow for such occurrences, there was a substantial reduction in the (already small) number of queries where the F4 weights were more effective (Robertson and Willett, 1996). In general then, the F4 weights gave a level of performance that was only marginally inferior to those provided by the GA weights, which had been obtained by a detailed exploration of the term-weight space defined by the query terms. This being so, it seems reasonable to conclude that the F4 weights give a practicable upperbound to the performance that is achievable in relevance feedback searches of text databases.

The GA that was used for these relevance feedback experiments was also designed to calculate weights for *substructural analysis*, (Cramer *et al.*, 1974). Here, weights are calculated that relate the presence of a molecular feature to the probability that that molecule is active in some biological test system (*cf* relating the presence of a specific index term in a document to the probability that that document is relevant to a particular query). Given some training set of compounds for which the biological activities are available, the aim of substructural analysis is to develop weights that can then be used to select new compounds for biological testing. Specifically the sum of weights is calculated for the fragment substructures present in a molecule, and then the compounds are ranked in order of decreasing sums-of-weights, so that chemical synthesis and biological assays can be focused on those compounds that have a high *a priori* probability of activity.

The use of substructural analysis methods with fragment substructures (such as those discussed in the second section of this paper) is well established, with many different types of weighting scheme having been described in the literature (Ormerod *et al.*, 1989). The project to be described here used a rather different level of molecular description, specifically high-level molecular characteristics suggested by medicinal chemists at GlaxoWellcome Research and Development (our collaborators in this project) as affecting the drug-like behaviour of molecules. The features are the distribution of property values for the molecular weight, the ^{2}k a shape index (Kier, 1987), and the numbers of aromatic rings, rotatable bonds, hydrogen-bond donor atoms and hydrogen-bond acceptor atoms, in the molecules comprising a database. Each property was allocated a total of 20 bins; for example, the first bin for the hydrogen-bond donor feature describes those molecules in a dataset that have no donor atoms, the second bin those that have one donor atom, and so in until the 20th and last, which represents those that have 19 or more donor atoms. The molecular weight and shape index bins contained ranges of values, rather than specific integer counts, *e.g.*, the first two bins for molecular weight described the molecules with weights in the range 0-74.99 and 75-149.99, *etc.*

Table 1. Retrieval effectiveness in relevance feedback searches using F4 and GA weights. [The Mean Recall portion of the table is the mean recall when averaged over all of the queries for a particular test collection using the two types of weight, while the right-hand portion shows the number of queries where the GA and F4 weights gave the same recall value (Same), and where either the F4 weights or the GA weights were better.]

Document	**Mean Recall**		**Number Of Queries**		
Collection	**F4**	**GA**	**Same**	**F4 Better**	**GA Better**
Keen	0.55	0.60	31	2	30
Cranfield	0.58	0.65	117	6	102
Harding	0.34	0.33	33	16	16
Evans	0.38	0.42	11	7	21
LISA	0.61	0.64	13	5	17
SMART	0.30	0.34	23	10	44
UKCIS	0.18	0.21	105	7	70

The GA described previously was used to calculate weights for each bin of each feature, with each such weight reflected the extent to which possession of that feature-value combination resulted in a molecule having some specific activity. The fitness function was then based on the occurrences of these feature-value pairs in sets of molecules for which activity data are available. For this purpose, we used sets of structures extracted from the *World Drugs Index* (or WDI, available from Derwent Information at URL http://www.derwent.co.uk) and *SPRESI* (available from Daylight Chemical Information Systems Inc. at http://www.daylight.com) databases: the former contains molecules for which biological activities have been established while the latter contains a large number of molecules that are assumed to be inactive. As with the text application, the chromosome in the GA encodes possible weights (in this case for each of the feature-value bins), and then the score for a particular molecule is the sum of the weights for those feature values that are associated with it. The fitness function of the GA is simply the number of top-ranked active molecules once the molecules have been ranked in decreasing order of these sums-of-weights.

Sets of 1000 molecules with some particular activity were chosen from the WDI and then combined with a set of 16,807 SPRESI structures. The GA was used to calculate weights for this combined dataset, and a note made of the number of actives in the top 1000 positions once convergence had occurred. The effectiveness of the weights is measured by the degree of *initial enhancement, i.e.*, the ratio of the observed number of top-1000 actives to the number of actives that would be obtained by selecting 1000 compounds at random (Gillet *et al.*, 1998). These results are summarised in Table 2 and it will be seen that the GA weights give consistently better results than merely selecting compounds at random for testing, although the extent of the improvement is clearly dependent upon the specific activity under investigation. Gillet *et al.* (1998) report comparable results from extended experiments using various forms of these weights on a range of datasets; these involved not only retrospective studies (as here) but also experiments where the weights were used in a predictive manner. It was concluded that the GA provided a simple and effective way of ranking sets of compounds in order of decreasing probability of activity, thus enabling the prioritisation of compounds for synthesis and biological testing.

Table 2. Effectiveness of ranking active compounds using the GA-based substructural analysis weights. [The Drug Activity listed is that contained in the keyword activity field for each of the compounds in the WDI database.]

Drug Activity	Initial Enhancement
Hormones	8.3
Anti-cancers	7.6
Anti-microbials	7.6
Anaesthetics	4.1
Blood	3.6
Central nervous system	3.9
Psychotropics	2.5

5. COMBINATION OF CHEMICAL SIMILARITY MEASURES USING DATA FUSION

The many types of similarity measures that are available for the measurement of molecular similarity has led to comparative studies in which researchers try to identify a single, "best" measure, using some quantitative performance criterion. For example, Willett *et al.* (1986) discuss the merits of different association coefficient for chemical similarity searching and conclude by recommending the Tanimoto coefficient for this purpose. Such comparisons, of which there are many in the chemical information literature, are limited in that they assume, normally implicitly, that there is some specific type of structural feature (similarity coefficient, weighting scheme or whatever it is that is being investigated) that is uniquely well suited to describing the type(s) of biological activity that are being sought for in a similarity search. The assumption cannot be expected to be generally valid, given the multi-faceted nature of biological activities, and this has led us to consider chemical applications of *data fusion* (Hall, 1992).

Data fusion was developed to combine inputs from different sensors, with the expectation that using multiple information sources enables more effective decisions to be made than if just a single sensor was to be employed. The methods are used in a wide range of military, surveillance, medical and production engineering applications (see, *e.g.*, Arabnia and Zhu (1998)): our interest was aroused by a paper by Belkin *et al.* (1995), in which data fusion was used to combine the results of different searches of a text database, conducted in response to a single query but employing different indexing and searching strategies. A query was processed using different strategies, each of which was used to ranking the database in order of decreasing similarity with the query. The ranks for each of the documents were then combined using one of several different fusion rules, the output of the fusion rule was taken as the document. s new similarity score and the fused lists were then re-ranked in descending order of similarity.

Ginn *et al.* (1999) have described the application of these ideas in the context of matching a target structure against a database, using several different measures of chemical similarity as summarised in Figure 1.

1) Execute a similarity search of a chemical database for some particular target structure using two, or more, different measures of inter-molecular similarity.

2) Note the rank position, r_i, of each database structure in the ranking resulting from use of the i-th similarity measure.

3) Combine the various rankings using a fusion rule to give a new combined score for each database structure

4) Rank the resulting combined scores, and then use this ranking to calculate a quantitative measure of the effectiveness of the search for the chosen target structure.

The fusion rules used were those identified by Belkin *et al.* (1995) and shown in Figure 2, and the combined scores output by the fusion rule are then used to re-order the database structures to give the final ranked output. It will be seen from Figure 2 that the MIN and MAX rules represent the assignment of extreme ranks to database structures and it is thus hardly surprising that both can be highly sensitive to the presence of a single "poor" similarity measure amongst those that are being combined. The SUM rule, where each database structure is assigned the sum of all the rank positions at which it occurs in the input lists, is expected to be more stable against the presence of a single poor or noisy input ranking, and this was generally found to be the rule of choice in our experiments (Ginn *et al.*, 1999).

1. Execute a similarity search of a chemical database for some particular target structure using two, or more, different measures of inter-molecular similarity.
2. Note the rank position, r_i, of each database structure in the ranking resulting from use of the i-th similarity measure.
3. Combine the various rankings using a fusion rule to give a new combined score for each database structure
4. Rank the resulting combined scores, and then use this ranking to calculate a quantitative measure of the effectiveness of the search for the chosen target structure.

Figure 1. Combination of similarity rankings using data fusion

Name	Fusion Rule
MIN	minimum (r_1, r_2 ,&r_i &r_n)
MAX	maximum (r_1, r_2,&r_i &r_n)
SUM	$\sum_{i=1}^{n} r_i$

Figure 2: Fusion rules for combining n ranked lists, where r_i denotes the rank position of a specific database structure in the i-th (1 £ i £ n) ranked list.

Our experiments involved combining searches of several different datasets using several different similarity measures in each case. The dataset considered here contained 75 compounds selected by Kahn (1998) in a discussion of various types of structural descriptor, with each of the compounds belonging to one of 14 well-defined biological activity classes (such as angiotensin-converting enzyme inhibitors and HIV-1 protease inhibitors). Six different types of similarity measures were used in the experiments, as detailed by Ginn *et al.* (1999); for the present, we note merely that they encoded information about the steric, electrostatic and hydrophobic characteristics of molecules (similarity measures denoted by the symbols "F" and "J"), about the 3D arrangement of pharmacophore points in molecules (denoted by the symbols "3" or "T") and about the occurrences of chains of up to 7 non-hydrogen atoms (denoted by the symbols "2" or "N"). The SUM rule was used to generate all possible combinations of rankings from these six similarity measures. Each of the members of the dataset was used as the target structure for a similarity search and Table 3 details the mean numbers of actives (*i.e.*, molecules with the same activity as the target structure) found in the top-10 nearest neighbours when averaged over all 75 searches. The values of *c* at the top of the table denote the number of similarity measures that were fused (so that, *e.g.*, *c*=1 represents the original measures and *c*=2 represents the fusion of a pair of the original measures) and a shaded element indicates a fused combination that is better than the best original individual measure (which was one of the 3D pharmacophore measures).

It will be seen that very many of the fused combinations in Figure 3 are shaded, thus supporting the idea that improvements in effectiveness can be achieved by using more than just a single similarity measure. The table also shows that the fraction of the combinations that are shaded increases in line with *c*, so that all combinations with c^3 4 perform at least as well as the best of the individual similarity measures. That said, the best result overall was obtained with 23FJT (rather than with 23FJNT, the combination involving all of the individual measures): thus, while simply fusing as many individual measures as are available appears to work very well, superior results may be obtained (for this dataset at least) from fusing a subset of the individual measures.

Other datasets studied by Ginn *et al.* (1999) were: 8178 molecules from the Starlist file (available from BioByte Corp. at URL http://clogp.pomona.edu) for which experimental values of the octanol/water partition coefficient, an important parameter in statistical methods for the prediction of biological activity, were available; several sets of 3,500 molecules from the WDI database for which biological activity data were available; and 136 biological dyes used to stain cells so as to visualise various organelles. In all cases, it was found that use of a fusion rule such as SUM generally resulted in a level of performance (however quantified) that was at least as good as (and often noticeably better than) the best individual measure. The best individual measure often varies from one target structure to another in an unpredictable manner, and the use of a fusion rule will thus generally provide a more consistent level of searching performance than will a single measure of chemical similarity. With the increasing number of such measures available that can be implemented with a reasonable degree of efficiency (Willett *et al.*, 1998), it would seem appropriate to consider the use of more than one of them for similarity searching of chemical structure databases.

6. CONCLUSIONS

This paper has presented several examples of algorithms and data structures that are applicable to the processing of both textual and chemical databases. These examples suggest that each area has much to offer to the other, with the transfer of methodology occurring from chemical to textual applications, and *vice versa*.

The substructure searching methods discussed in the second section provide an excellent example of this transfer of ideas. Here, an approach that was first developed for chemical retrieval (specifically in the context of 2D substructure searching) was soon shown to be applicable to many applications in the general area of textual databases. One such application, distribution sorting, resulted in an algorithm that was then adapted for use in a rather different area of chemical retrieval (3D substructure searching). In fact, although not discussed here, the idea of using distance and angular information for database searching has now been extended to the computationally demanding task of searching for patterns in the 3D structures of proteins, where our search methods have led to the discovery of many previously unknown structural relationships between proteins (see, *e.g.*, Artymiuk *et al.*, 1996, 1997). In the case of the hypertext-comparison project, the similarity measures discussed above were originally developed for calculating the similarities between pairs of 2D molecules, while the GA for the calculation of weights was designed from the start for use in both relevance feedback and substructural analysis. Finally, our use of data fusion for combining chemical similarity measures was a simple application of work done by others in the textual domain. Given the range of applications we have already been able to identify, we trust that other researchers will be encouraged to investigate the many similarities that exist between chemical and textual database processing.

Table 3. Mean number of actives found in the top-10 positions for chemical similarity searches when combining various numbers, *c*, of different similarity measures. [The pale-green shading indicates a fused result at least as good as the best original similarity measure.]

c=1		*c*=2		*c*=3		*c*=4		*c*=5		*c*=6	
2>	0.80	23	1.10	23F	1.28	23FJ	1.52	23FJN	1.45	23FJNT	1.43
3	1.12	2F	1.04	23J	1.39	23FN	1.23	23FJT	1.69		
F	0.89	2J	1.01	23N	1.04	23FT	1.43	23FNT	1.36		
J	1.08	2N	0.68	23T	1.24	23JN	1.31	23JNT	1.43		
N	0.63	2T	0.95	2FJ	1.35	23JT	1.45	2FJNT	1.43		
T	0.69	3F	1.09	2FN	1.08	23NT	1.25	3FJNT	1.51		
		3J	1.25	2FT	1.28	2FJN	1.28				
		3N	1.00	2JN	1.03	2FJT	1.53				
		3T	1.32	2JT	1.10	2FNT	1.28				
		FJ	1.20	2NT	0.95	2JNT	1.17				
		FN	0.91	3FJ	1.40	3FJN	1.35				
		FT	1.11	3FN	1.19	3FJT	1.55				
		JN	0.89	3FT	1.33	3FNT	1.41				
		JT	0.93	3JN	1.25	3JNT	1.36				
		NT	0.85	3JT	1.45	FJNT	1.32				
				3NT	1.20						
				FJN	1.11						
				FJT	1.21						
				FNT	1.11						
				JNT	1.12						

ACKNOWLEDGEMENTS.

I thank all of my colleagues, past and present, for their contributions to the research that has been summarised here, and the many organisations that have provided financial support for my research over the years. The Krebs Institute for Biomolecular Research is a designated Biomolecular Sciences Centre of the Biotechnology and Biological Sciences Research Council.

REFERENCES

1. Adamson, G.W., Cowell, J., Lynch, M.F., McLure, A.H.W., Town, W.G. and Yapp, A.M. (1973) "Strategic considerations in the design of screening systems for substructure searches of chemical structure files." *Journal of Chemical Documentation*, **13**, 153-157.

2. Arabnia, H.R. and Zhu, D., editors (1998). *Proceedings of the International Conference on Multisource-Multisensor Information Fusion, Fusion. 98*. CSREA Press.

3. Artymiuk, P.J., Poirrette, A.R., Rice, D.W. and Willett, P. (1996) "Biotin carboxylase comes into the fold." *Nature Structural Biology*, **3**, 128-132.

4. Artymiuk, P.J., Poirrette, A.R., Rice, D.W. and Willett, P. (1997) "A polymerase I palm in adenylyl cyclase?" *Nature*, **388**, 1997, 33-34.

5. Ash, J.E., Warr, W.A. and Willett, P., editors (1991) *Chemical Structure Systems*. Chichester: Ellis Horwood.

6. Barnard, J.M. (1993) "Substructure searching methods: old and new." *Journal of Chemical Information and Computer Sciences*, **33**, 532 - 538.

7. Barnard, J.M., Lynch, M.F. and Welford, S.M. (1984) "Computer storage and retrieval of generic chemical structures in patents. Part 6. An interpreter program for the generic structure language

GENSAL." *Journal of Chemical Information and Computer Sciences*, **24**, 66-70.

8. Barton, I.J., Creasey, S.E., Lynch, M.F. and Snell, M.J. (1974) "An information-theoretic approach to text-searching in direct-access systems." *Communications of the Association for Computing Machinery*, **17**, 345-350.

9. Belkin, N.J., Kantor, P., Fox, E.A. and Shaw, J.B. (1995) "Combining the evidence of multiple query representations for information retrieval." *Information Processing and Management*, **31**, 431-448.

10. Carhart, R.E., Smith, D.H., Venkataraghavan, R. (1985) "Atom pairs as molecular features in structure-activity studies: definition and applications." *Journal of Chemical Information and Computer Sciences*, *25*, 64-73.

11. Cooper, D., Dicker, M.E. and Lynch, M.F. (1980) "Sorting of textual databases: a variety generation approach to distribution sorting." *Information Processing and Management*, **16**, 49-56.

12. Cooper, D. and Lynch, M.F. (1984) "The use of binary search trees in external distribution sorting." *Information Processing and Management*, **20**, 547-557.

13. Cramer, R.D., Redl, G. and Berkoff, C.E. (1974) "Substructural analysis. A novel approach to the problem of drug design." *Journal of Medicinal Chemistry*, **17**, 533-535.

14. Crestani, F. and Melucci, M. (1998) "A case study of automatic authoring: From a textbook to a hyper-textbook." *Data and Knowledge Engineering*, **27**, 1-30.

15. Cringean, J.K., Pepperrell, C.A., Poirrette, A.R. and Willett, P. (1990) "Selection of screens for three-dimensional substructure searching." *Tetrahedron Computer Methodology*, **3**, 37-46.

16. Edgar, S.J., Holliday, J.D. and Willett, P. (1999) "Effectiveness of retrieval in similarity searches of chemical databases: a review of performance measures." In preparation.

17. Ellis, D., Furner-Hines, J. and Willett, P. (1994) "On the creation of hypertext links in full-text documents: Measurement of inter-linker consistency." *Journal of Documentation*, **50**, 67-98.

18. Ellis, D., Furner-Hines, J. and Willett, P. (1996) "On the creation of hypertext links in full-text documents: Measurement of retrieval effectiveness." *Journal of the American Society for Information Science*, **47**, 287-300.

19. Gillet, V.J., Willett, P. & Bradshaw, J. (1998) "Identification of biological activity profiles using substructural analysis and genetic algorithms." *Journal of Chemical Information and Computer Sciences*, **38**, 165-179.

20. Ginn, C.M.R., Willett, P. and Bradshaw, J. (1999) "Combination of molecular similarity measures using data fusion." *Perspectives in Drug Discovery and Design*, submitted for publication.

21. Hall, D.L. (1992) *Mathematical Techniques in Multisensor Data Fusion*. Northwood, MA: Artech House.

22. Jakes, S.E. and Willett, P. (1986) "Pharmacophoric pattern matching in files of 3-D chemical structures: selection of inter-atomic distance screens." *Journal of Molecular Graphics*, **4**, 12-20.

23. Kahn, S.D. (1998) "Combinatorial libraries: structure activity analysis, in: Schleyer, P.v.R., Allinger, N.L., Clark, T., Gasteiger, J., Kollman, P.A., Schaefer III, H.F. and Schreiner, P.R., editors, *Encyclopedia of Computational Chemistry*. Chichester: John Wiley. Vol 1, pp. 417-425.

24. Kier, L.B. (1987) "Indexes of molecular shape from chemical graphs." *Medicinal Research Reviews*, **7**, 417-440.

25. Lynch, M.F. (1977) "Variety generation - a re-interpretation of Shannon. s mathematical theory of communication and its implications for information science." *Journal of the American Society for Information Science*, **28**, 19-25.

26. Lynch, M.F. and Willett, P. (1987) "Information retrieval research in the Department of Information Studies, University of Sheffield." *Journal of Information Science*,**13**, 221-234.

27. Markey, K. (1984) "Inter-indexer consistency tests: A literature review and report of a test of consistency in indexing visual materials." *Library and Information Science Research*, **6**, 155-177.

28. Martin. Y.C. and Willett, P., editors (1998) *Designing Bioactive Molecules: Three-Dimensional Techniques and Applications*. Washington: American Chemical Society.

29. Maybury, M.T., editor (1997) *Intelligent Multimedia Information Retrieval*. Cambridge MA: MIT Press.

30. Ormerod, A., Willett, P. and Bawden, D. (1989) "Comparison of fragment weighting schemes for substructural analysis." *Quantitative Structure-Activity Relationships*,**8**, 115-129.

31. Robertson, A.M. and Willett, P. (1996) "An upperbound to the performance of ranked-output searching: optimal weighting of query terms using a genetic algorithm."*Journal of Documentation*, **52**, 1996, 405-420.

32. Robertson, S.E. and Sparck Jones, K. (1976) "Relevance weighting of search terms." *Journal of the American Society for Information Science*, **27**, 129-146.

33. Salton, G. (1989) *Automatic Text Processing*. Reading, MA: Addison-Wesley.

34. Salton, G. and Buckley, C. (1990) "Improving retrieval performance by relevance feedback." *Journal of the American Society for Information Science*, **41**, 288-297.

35. Sparck Jones, K. and Willett, P., editors (1997) *Readings in Information Retrieval*. San Francisco: Morgan Kaufmann.

36. Welford, S.M., Lynch, M.F. and Barnard, J.M. (1981) "Computer storage and retrieval of generic chemical structures in patents. Part

3. Chemical structure grammars and their role in the manipulation of chemical structures." *Journal of Chemical Information and Computer Sciences*, **21**, 161-168.

37. Willett, P. (1981) "A fast procedure for the calculation of similarity coefficients in automatic classification." *Information Processing and Management*, **17**, 53-60.

38. Willett, P. (1982) "The calculation of inter-molecular similarity coefficients using an inverted file algorithm." *Analytica Chimica Acta*, **138**, 339-342.

39. Willett, P. (1995) "Searching for pharmacophoric patterns in databases of three-dimensional chemical structures." *Journal of Molecular Recognition*, **8**, 290-303.

40. Willett, P. (1997) "Information retrieval research in the University of Sheffield", *ACM SIGIR Forum*, **31**(2), 7-13.

41. Willett, P., Barnard, J.M. and Downs, G.M. (1998) "Chemical similarity searching." *Journal of Chemical Information and Computer Sciences*, **38**, 983-996.

42. Willett, P., Winterman, V., Bawden, D. (1986) "Implementation of nearest neighbour searching in an online chemical structure search system **26**, 36-41.

9

INFORMATION AND KNOWLEDGE MANAGEMENT: DIMENSIONS AND APPROACHES

Christian Schlögl

Institute of Information Science, University of Graz, Graz, Austria

ABSTRACT

Introduction

Though literature on information and knowledge management is vast, there is much confusion concerning the meaning of these terms. Hence, this article should give some orientation and work out the main aspects of information and knowledge management.

Method

An author co-citation analysis, which identified the main dimensions of information management, forms the basis of the study of literature. In it, the main aspects of information management are further refined. Furthermore, it will be investigated if the concept of knowledge management adds anything to information management, and if so what it is.

Analysis

Data for analysis were retrieved from Science Citation Index and Social Science Citation Index. Though the literature review is based on the quantitative results of the bibliometric analysis and has tried to consider the most prominent publications, some degree of subjectivism cannot be excluded.

Results

As a result of analysis, a distinction can be made between content and technology-oriented information management approaches. According to the literature review, technology-oriented information management includes data management, information technology management and strategic information technology management. The main emphasis of these approaches is the effective and efficient use of information technology. In contrast, content-oriented approaches focus on information and its use. They can be distinguished in records management, provision of external information, human-centered information management, and information resources management. The reading of the literature on knowledge management reveals, that this term is either used synonymously for information management or for the *management of work practices* with the goal of improving the generation of new knowledge and the sharing of existing knowledge.

Conclusions

This article identifies various aspects that are embraced by the terms *information management* and *knowledge management*. Thus, it should contribute to more terminological clarity and finally improve communication both in science and in professional practice.

1. INTRODUCTION

Information management has been around for more than two decades. Many authors date its beginning back to the Paperwork Reduction Act of 1980 in which U.S. federal agencies were forced to introduce information resource management. Regardless of its exact origins, there was a substantial growth in literature dealing with this topic at the beginning of the eighties. In the second half of the nineties, the term knowledge management became more popular.

Though literature on information and knowledge management is expansive, there is much confusion concerning the meaning of these terms. This is the starting point of this article the aim of which is to identify the main dimensions and major approaches of information and knowledge management. This should also give answers to the question, 'Does the concept of knowledge management add anything and, if so, what is it?'

2. MAPPING THE LITERATURE OF INFORMATION MANAGEMENT

In order to cope with the voluminous literature on information management, a bibliometric study was performed (Schlogl 2003). By means of an author co-citation analysis of data from Science Citation Index and Social Science Citation Index, the intellectual structure of information management was mapped (see Figure 1).

In order to select the authors, all publications including either the phrase *information management* or *information resource(s) management* in the title were retrieved. The following DIALOG command was used: SELECT ((INFORMATION () MANAGEMENT (NOT W) SYSTEM?) OR (INFORMATION () RESOURCE? () MANAGEMENT)) /ART /TI. The authors who received most citations from these information management articles constituted the basis for co-citation analysis. (Problems which are associated with bibliometric studies in general and with a co-citation analysis covering multiple disciplines in particular are discussed extensively by Schloegl (2000).) For mapping the authors, a statistical method called multidimensional scaling was used. The calculations were done by means of the ALSCAL program which is included in SPSS.

Multidimensional scaling locates the authors in two or more dimensional spaces based on their similarity (number of co-citation counts). The arrangement of the authors is made as follows:

- authors who are co-cited many times are grouped close together.

- authors with many links to others tend to be in a central position, while authors with only a few links will be placed more peripherally.

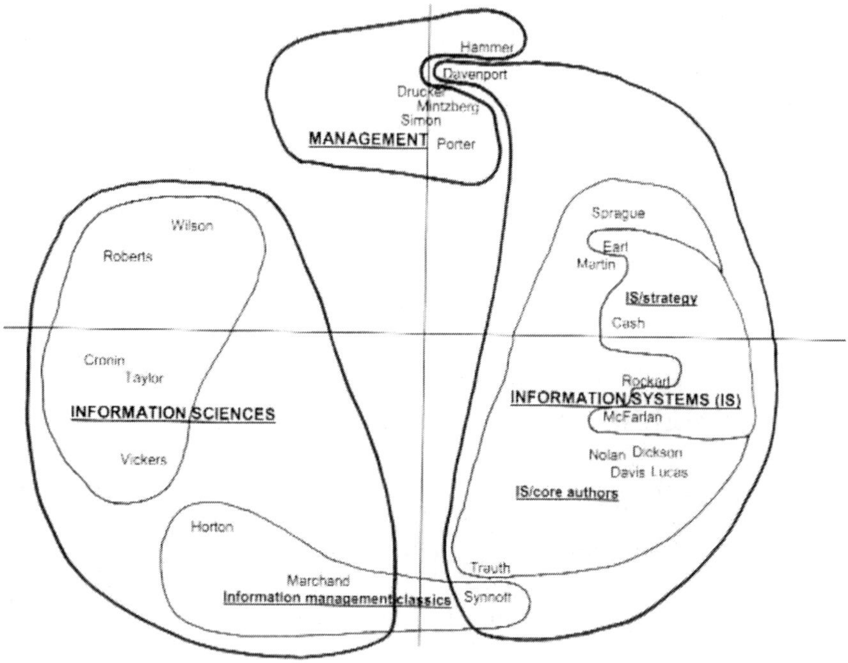

Figure 1. Map of information management (source: Schlögl 2003)

Besides providing an information-rich display of the co-citation links, the second main use of multidimensional scaling is to identify the salient dimensions underlying their placement (McCain 1990). On the vertical axis this is the extent to which the authors worked on the topic *information management*. On the top, there are scholars from management science and business administration. These authors did not focus their research on information management itself, but others used their work as a foundation, especially in the field of information systems. On the bottom, there are those authors whose work concentrated primarily on information management. These authors published the first books on this topic. They can be referred to as the *classical* writers on information management.

The horizontal axis indicates the different subject dimensions. Authors on the left can be assigned to the information sciences. Their research interest lies in information and its use. On the right, there are those authors whose research topic is the efficient and effective use of information technology (IT). The related disciplines are information systems and business informatics.

Interestingly, no authors were placed in the centre of the map. This might suggest that information management is not an interdisciplinary but (only) a multidisciplinary topic, especially as there is no collaboration between researchers from information systems and those from information sciences. Ellis *et al.*(1999) arrive at a similar conclusion after having analysed the literature on information retrieval and user studies. They conclude that there is a 'lack of contact' between these two disciplines.

According to these interpretations, a distinction can be made between content-oriented and technology-oriented information management. In the following sections, various approaches will be discussed for each of these two dimensions of information management. Then the concept of knowledge management will be investigated, in particular its relationship to information management.

3. TECHNOLOGY-ORIENTED INFORMATION MANAGEMENT

The general purpose of information management is to make available the right information at the right time and at the right place. For technology-oriented information management, computer-based information systems are the primary means to this end. Information management from this angle stresses the importance of information technologies. This is justified by the high degree of IT integration, the complexity of its application, and its strong consequences for an organization.

Within technology-oriented information management, the following aspects can be distinguished:

- data management,
- IT management, and

- strategic use of IT.

4. DATA MANAGEMENT

Some authors equate information management with data management (e.g., Hoven 1995). This is due to the high importance of data. Data is needed for and produced by nearly every activity and is an important input for almost all decisions on each level of an enterprise (Levitin & Redman 1998).

In a broader sense, data management can be defined as all organizational and technical tasks concerning the planning, storage, and provision of data, both for computer personnel and end-users (Schulte 1987). Its goal is to maximize the quality, usability, and value of the data resources in an enterprise (Hoven 1995). In the literature, often two components are suggested for data management: data administration and database administration (Gilllenson 1985; Henderson 1987;Lytle 1986). Data administration primarily serves a planning and analysis function. It may be responsible for data planning, accountability, policy development, standards setting, and support. One of the major tasks includes the design of the data architecture of an organization. Database administration provides a framework for managing the data on an operational level. Its role may include performance monitoring, troubleshooting, security monitoring, physical database design, and data backup.

5. IT MANAGEMENT

There is a broader view of information management that data management is a part of or a precondition for information management (English 1996; Heinrich 2002; Henderson 1987; Schwarze 1988). The management of hardware, software, and IT personnel must be included as well. The emphasis here is on the technological aspects of electronic data processing.

One well-structured example for these approaches was suggested by Wollnik (1988). His model distinguishes three levels (see Figure 2): the management of information use (upper level), the management of information systems (middle level), and the management of the information infrastructure (lower level).

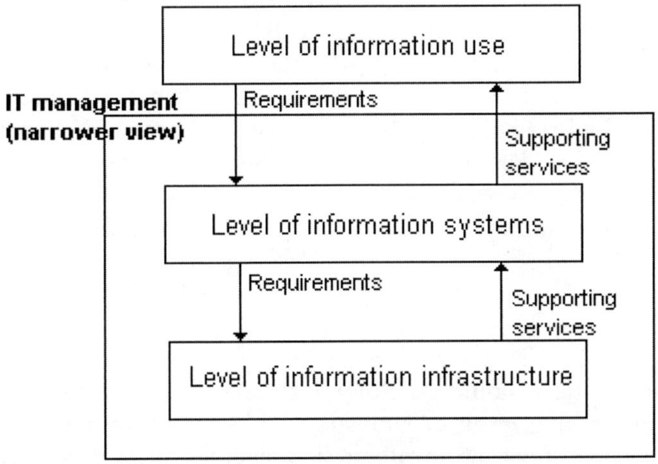

Figure 2. Three-level model of information management (source: Wollnik 1988)

Information infrastructures provide means for all possible types of data processing open to different use (e.g., computers, networks). They constitute the basis for the information systems that, in contrast, support specific tasks of an organization. The information systems provide, in particular, the means for information use and exchange. The main role of information systems management is the management of the development and operation of an organization's information systems. Information management from this view can be defined as the planning, organization, and control of information use, information systems, and information infrastructure in an organization.

6. STRATEGIC USE OF INFORMATION TECHNOLOGY

The use of information technology as a strategic resource has attracted much attention especially in Anglo-American countries in the past. Many publications on this topic deal with the strategic relevance of information processing. These publications explore the question whether and to what extent information technology can contribute to the objectives of an organization (Krueger & Pfeiffer 1988; McFarlan 1984; Porter & Millar 1985). Some authors were concerned with models to support strategic applications (Ives & Learmonth 1984; Porter & Millar 1985; Rockart & Crescenzi 1984; Wiseman 1985). A comprehensive survey about examples for the competitive use of information technology is given by Mertens and Plattfaut (1986) for instance. While the first publications were very "enthusiastic", there has been some disillusionment in latter years. Several authors are now concerned with the question of whether an IT-based competitive advantage can be maintained for a longer time (Hatten & Hatten 1997; Kettinger, *et al.* 1994; Mata, *et al.* 1995; Miron, *et al.* 1988; Ross, *et al.* 1996). Some authors doubt whether this is possible at all (e.g., Sutherland 1991).

A strategic orientation should go along with an increased valuation of information processing in an organisation. This should be met by the creation of a new post at strategic management level: the chief information officer (Synnott 1987b). The main purpose behind this is to ensure that information processing is coordinated with corporate objectives.

Strategic aspects play an important role in nearly all concepts that are concerned with the management of information technology. For instance, the title of the book by Synnott (1987a) indicates the significance of strategic aspects in his approach. This applies to O'Brien and Morgan (1991) as well as to Lewis and Martin (1989) who view strategy as a fundamental characteristic of information [technology] management. Pietsch, *et al.* (1998) have even called their approach*strategic information [technology] management.*

7. CONTENT-ORIENTED INFORMATION MANAGEMENT

Authors in this area of information management usually have a background in library and information science, records management, or a closely related discipline. Contrary to technology-oriented information management, these approaches focus on information content.

Publications can be classified here as follows:

- records management,
- provision of external information,
- human-centred information management, and
- information resources management.

8. RECORDS MANAGEMENT

According to Savic (1992) and Trauth (1989), records management is one of the oldest information management disciplines. This claim was renewed by the records management community when renaming the most important international journal in the field from *Records Management Quarterly* to *The Information Management Journal*. In a highly respected book in this field, records management is described as a discipline which is concerned with the management of document-based information systems (Robek, *et al.* 1996).

The main goals of records management are:

- to furnish accurate, timely, and complete information in order to enable efficient decision making processes;
- to process recorded information as efficiently as possible;
- to provide information and documents at the lowest cost;
- to render the maximum utility to the users of documents;
- to dispose of records which are no longer needed (Robek, *et al.* 1996).

As can be seen, the main feature of records management is the management of the information life cycle. According to Robek, *et al.* (1996), it consists of production, dissemination and use, storage and provision for current access, decisions on the retention/destruction, and archiving of documents.

Contrary to data management, the emphasis of records management lies in textual, document-based information (memos, reports, letters, etc.) (Noot 1998) originally recorded on paper. Nevertheless recent developments in information technologies have caused significant changes in records management. Correspondingly, the notion of a document acquired a wider meaning. Records management deals less and less with documents recorded on paper. The management of electronic documents has become more central instead using support systems like intranets, electronic imaging systems, and workflow management systems.

9. PROVISION OF EXTERNAL INFORMATION

Content-oriented information management has a much stronger focus on the provision of external information (Huebner 1996). In some publications only the provision of external information (from databases) (Kind 1986; Meik 1997) or information and documentation (Kroll 1990) are related to information management. For other authors, the provision of external information is an important part of information management (e.g. Gazdar 1989; Kmuche 1997).

Kuhlen and Finke (1988) claim that external information concerning changes in relevant segments of the environment is much more important for the success of an organization than the management of information technology. Choo (1998) has similar arguments. According to him, the survival of a company depends on how well it processes information about its environment and, as a result, succeeds in adapting efficiently to environmental changes. This also shows the importance of external information for strategic planning (Anthony 1988).

10. HUMAN-CENTERED INFORMATION MANAGEMENT

Humans play a much more important role in content-oriented management approaches (Grudowski 1995; Taylor 1986; Wang 1998; Wersig 1989; Zijlker 1988). For instance, Wersig's approach to information management does not centre on information technology or formal theories but on how humans handle information in reality

(Grudowski 1995). While it is typical for technology-oriented information management approaches to model information systems in a very formal way (entity relationship diagrams, data flow diagrams, etc.), content-oriented concepts use much less formalistic methods (e.g., Cronin & Davenport 1991; Wersig 1989). This has also to do with the fact that the term information system has a much wider meaning, which is not only restricted to computers (Bergeron 1996). For this reason, it is not the goal to thoroughly automate and completely formalize information processes (Cronin & Davenport 1991). According to Wersig (1989), order should only be established for circumstances where it is essential; allowing creative solutions with their inherent qualities to unfold from what may appear to be chaos.

Some approaches do not only consider information handling and/or information behaviour of individuals but of the organization as a whole. They also include an information culture that according to Davenport (1997) results from the total information behaviour of the organization's staff. Schneider (1990) even calls the approach developed by her 'culture-conscious information management'. It takes into consideration that information management is embedded into a certain organization with a specific history and based on general assumptions that influence the behaviour of its staff.

11. INFORMATION RESOURCES MANAGEMENT

A series of authors mainly from the information sciences take a holistic perspective on information management integrating more or less all the aspects discussed above. Although there is not a well established term, these approaches are referred to as information resources management in the following. According to Bergeron (1996), IRM is grounded in the following assumptions:

- recognition of information as a resource;
- an integrative management perspective;
- management of the information life cycle;
- a link with strategic planning.

One important feature of information resources management consists is that it is a framework that seeks to integrate different information professionals and functions under one umbrella (Bergeron 1996; Levitan 1982; Lytle 1986; Miller 1988; Otremba 1987; Vickers 1985; Wiggins 1988; Wilson 2002). However there are different points of view on which resources are to be managed (Vogel 1992). In a wide sense, information resources management can be defined as the management of those resources (human and physical) that are concerned with the systems support (development, enhancement, maintenance) and the servicing (processing, transformation, distribution, storage, and retrieval) of information (Schneyman 1985). Marchand and Horton (1986) distinguish between information resources and information assets. Information resources include information specialists; information technology; facilities like the library, the data processing department, and the information centre; and external information brokers. Information assets cover all the formal information holdings of an organization (data, documents, technical literature), know-how (rights on intellectual property, practical experience of staff), as well as knowledge about the environment (competitors; political, economic, and social environment). While the information assets concern information itself, the information resources are a means by which information can be gained. Lytle (1988) has a somewhat narrower view. He differentiates between data, hardware, software, information systems and services, as well as personnel.

12. KNOWLEDGE MANAGEMENT

Since the mid-nineties the label *knowledge management* has attracted much attention (Disterer, *et al.* 2002; Ponzi & Koenig 2002) while information management has been used less. As with information management, there is no agreement on what constitutes knowledge management (Corrall 1998; Morrow 2001; Wilson 2002). According to Wilson (2002), knowledge management is either used as a synonym for information management or for the 'management of work practices' which are to improve the sharing of knowledge in an organization.

The first perspective includes both content and technology-oriented information management. In the case of IT driven approaches there is a stronger focus on unstructured data (Davenport & Prusak 1998; Hoven 2001) and on different application areas of technology. They typically fall into the categories of knowledge databases and repositories (e.g., expertise of the after-sales staff), knowledge route maps and directories (e.g. yellow pages), and knowledge networks and communication tools (e.g., chat facilities, groupware) (Corrall 1998). Furthermore there is a shift in the underlying technology. Maurer (2003) distinguishes a knowledge management system from a conventional information system in that it has the following additional features: (i) it makes accessible private information that is of no use anymore to its creator but probably for others (e.g., former project documentation makes it possible to retrace decision processes), (ii) it learns from the users' use (e.g., many users who search x seek for y as well), (iii) it can initiate actions and provide information without any request by the user (e.g., if a user retrieves x, y is offered to him automatically), (iv) it can generate new information from existing (e.g., automatic classification).

One main shortcoming of the technologist's view is that it assumes that knowledge can be codified which is not usually the case (Ponzi 2002). Though technology can play an important role, knowledge management is not about technology (Al-Hawamdeh 2002). This leads to the discussion of the innovative aspects of knowledge management. But, first, the meaning of the term *knowledge* needs explaining. This article follows the definition by Wilson who defines knowledge as what a person knows. He states that, 'knowledge involves the mental processes of comprehension, understanding and learning that go on in the mind and only in the mind' (Wilson 2002). By contrast information is bound to a medium (which exists outside the mind). Before information can become knowledge, it must be incorporated into the knowledge structure of the particular human. This makes clear that knowledge cannot be managed. Rather, knowledge management is 'concerned with the management of work practices' (Wilson 2002) with the goal of improving the sharing of information (and, thereby, knowledge) in an organization. In line with this, knowledge management means the installation of purposeful management processes that enable to capture personal and contextual knowledge (Broadbent 1998) and thus permanently make accessible the staff's creativity for an organization (Scheid 2001). One of the most crucial factors is the establishment of a knowledge-friendly culture (Davenport, *et al.* 1998). Especially with regards to knowledge creation and transfer, flexible structures like personal networks or communities of practice (e.g., Brown & Duguid 1998) are important. The high claims of knowledge management can only be realized, if at all, after a long learning process. Therefore, organizational learning (e.g., Argyris 1991; Garvin 1993; Senge 1990) is an integral part of knowledge management.

As can be seen from what was mentioned above, knowledge management is predominantly an organizational sciences construct (Ponzi 2002). However, knowledge management has also a strong link to business strategy (Broadbent 1998), in that its proponents claim that it should enable an organization to cope effectively with rapid environmental change and to attain some form of competitive advantage (Morrow 2001).

Intellectual capital management is a closely related concept that is based on the view that in the information era the real market value of a company does not consist only of its physical and financial assets (book value) but more and more of its intangible assets. According to Corrall (1998) the choice between intellectual capital management and knowledge management depends on the emphasis given to measuring or managing knowledge assets. There were several models developed to measure intellectual capital (e.g. Edvinsson & Malone 1997; Kaplan & Norton 1992; Stewart 1997; Sveiby 1997). Basically, these models classify the intangible *capital* into the categories "customer capital" (e.g. customer relations, reputation, brands), "structural capital" (patents, business processes, culture, documented experience), and "human capital" (staff with their skills and experiences) (Ponzi 2002). As with knowledge management, the capabilities of the staff, especially their knowledge and experiences (human capital), should be converted to structural [and customer] capital and thus avoid a loss of the corporate memory (Corrall 1998).

13. CONCLUSIONS

This article has identified three major categories in the literature on information and knowledge management. They were referred to as technology-oriented information management, content-oriented information management, and knowledge management.

From a narrow point of view, data management is equated with information management. Even though information management is perceived as more comprehensive by nearly all authors, data management is one essential component of technology-oriented information management. IT management comprises of the planning, organization, and control of those tasks that are necessary for the provision and use of the information infrastructure as well as the (computer-based) information systems that are the basis for the information exchange in an organization. If possible the competitiveness of a company should be improved or at least its deterioration be prevented.

Content-oriented approaches are primarily concerned with the management of (codified) information. This relates both to the management of the information life cycle of mainly internally produced information and the provision of external information. Usually, they consider the user much more than technology-oriented approaches. Several authors who directed a particularly strong attention to humans were therefore attributed to their own sub-category. Strategic aspects are not dealt with as a central issue but are not of minor importance either. The emphasis is on external information and not on the competitive use of information technology. Information resources management has finally tried to integrate all the different aspects including the management of information technology creating a convergent view of information management.

If one does not consider its synonymous use for information management, knowledge management means the management of those work practices that aim at improving the generation of new and the sharing of existing knowledge. Some aspects of knowledge management were already partly dealt with by some authors of content-oriented approaches. However, the presented concepts were much less elaborated in this respect and mainly focused on behavioral aspects of information use. Knowledge management has a wider perspective working toward raising the creativity of the organization's staff. A few authors even developed some models for measuring the intellectual capital of an organization.

Table 1 summarizes the major aspects of information and knowledge management. It is mainly intended to serve as an orientation aid for the vast literature on this topic. Furthermore it should give a rough assessment of which disciplines (and professions) are engaged in which aspects of information and knowledge management. Accordingly, information systems and business informatics can be attributed to the management of information technology. Organizational and management sciences deal primarily with knowledge management (in a narrow sense). The management of (codified) information and the study of information use are the domains of records management and library and information science.

Table 1. Information and knowledge management: objects, terms and related disciplines

<table>
<tr><th colspan="2" rowspan="2">Objects</th><th colspan="2">Terms</th><th rowspan="2">Disciplines</th></tr>
<tr><th>Narrower terms</th><th>Broader terms</th></tr>
<tr><td rowspan="3">information technology</td><td>data (structure)</td><td>data management</td><td rowspan="3">IT management (technology-oriented information management)</td><td rowspan="3">information systems

business informatics</td></tr>
<tr><td>information system</td><td>information systems management</td></tr>
<tr><td>information infrastructure</td><td>management of information infrastructure</td></tr>
<tr><td rowspan="2">(codified) information</td><td>internal</td><td>records management</td><td rowspan="2">(content-oriented) information management</td><td rowspan="2">records management

library and information science</td></tr>
<tr><td>external</td><td>provision of external information</td></tr>
<tr><td colspan="2">work practices that relate to knowledge generation and sharing</td><td></td><td rowspan="2">knowledge management</td><td rowspan="2">organizational sciences

management sciences</td></tr>
<tr><td colspan="2">intellectual assets</td><td>intellectual capital management</td></tr>
</table>

As already mentioned, the terms *information management* and *knowledge management* are used very inconsistently in theory and in practice. This is due to ignorance and *tactical* reasons. The latter holds true for consultants and IT firms trying to market often *old* services and products under a new label (Wilson 2002). However, this is also partly valid for science. In a scientometric analysis, Mertens (1995) demonstrated how much business informatics suffers from emerging fashions that often go along with unnecessary changes of terms. For instance, this applies to issues like know-how databases and business intelligence which are investigated once again within the scope of knowledge management. Such scholarly behaviour could threaten a continuous accumulation of knowledge in the particular field in the long run. In this sense, the author hopes that this article will not only contribute to more terminological clarity but have further positive implications.

ACKNOWLEDGEMENT

The author would like to thank Mr. Norbert Berger from the University of Graz and Ms. Sheila French from Manchester Metropolitan University for proof-reading, Mr. Gunter Bauer for the technical assistance when creating this Web document, and the referees for their constructive comments.

REFERENCES

1. Al-Hawamdeh, S. (2002). Knowledge management: re-thinking information management and facing the challenge of managing tacit knowledge. *Information Research*,**8**(1). Retrieved 8 March, 2005 from http://InformationR.net/ir/8-1/paper143.html.

2. Anthony, R. (1988). *The management control function*. Boston, MA: Harvard Business School Press.

3. Argyris, C. (1991). Teaching smart people how to learn. *Harvard Business Review,* **69**(3), 99-109.

4. Bergeron, P. (1996). Information resources management. *Annual Review of Information Science and Technology*, **31**, 263-300.

5. Broadbent, M. (1998, May). The phenomenon of knowledge management: what does it mean to the information profession? *Information Outlook*, 23-31. Retrieved 20 May, 2005 from http://www.sla.org/pubs/serial/io/1998/may98/broadben.html

6. Brown, J. & Duguid, P. (1998). Organizing knowledge. *California Management Review*, **40**(3), 90-111.

7. Choo, C. W. (1998). *Information management for the intelligent organisation: the art of scanning the environment.* Medford, NJ: Information Today Inc.

8. Corrall, S. (1998). Knowledge management: are we in the knowledge management business? *Ariadne,* (18). Retrieved 8

March, 2005 from http://www.ariadne.ac.uk/issue18/knowledge-mgt/

9. Cronin, B. & Davenport, E. (1991). *Elements of information management.* Metuchen, NJ: Scarecrow Press

10. Davenport, T. (1997). *Information ecology: mastering the information and knowledge environment.* New York, NY: Oxford University Press.

11. Davenport, T.H., De Long, D.W. & Beers, M. C. (1998). Successful knowledge management projects. *Sloan Management Review,* **39**(2), 43-57.

12. Davenport, T. & Prusak, L. (1998). *Working knowledge: how organizations manage what they know.* Boston, MA: Harvard Business School Press.

13. Disterer, G., Krystofiak, S. & Blitzer, M. (2002). Informationen zum Wissensmanagement im Internet. *Wirtschaftsinformatik,* **45**(5), 547-554.

14. Edvinsson, L. & Malone, M. (1997). >*Intellectual capital: realizing your company's true value by finding its hidden brainpower.* New York, NY: HarperBusiness.

15. Ellis, D., Allen, D., & Wilson, T.D. (1999). Information science and information systems: conjunct subjects, disjunct disciplines. *Journal of the American Society for Information Science and Technology,* **50**(12), 1095-1107.

16. English, L.P. (1996). Redefining information management. *Information Systems Management,* **13**(1), 65-67.

17. Garvin, D. (1993). Building a learning organization. *Harvard Business Review,* **71**(4), 78-91.

18. Gazdar, K. (1989). *Informationsmanagement fuer Fuehrungskraefte: Konkrete Perspektiven fuer Wirtschaft, Verwaltung und Politik.* Frankfurt: Frankfurter Allgemeine Zeitung.

19. Gillenson, M. (1985). Trends in data administration. *MIS Quarterly,* **9**(4), 317-325.

20. Grudowski, S. (1995). *Informationsmanagement und Unternehmenskultur.* Unpublished PhD dissertaton, Freie Universitaet Berlin.

21. Hatten, M.L. & Hatten, K.J. (1997). Information systems strategy: long overdue - and still not here. *Long Range Planning,* **30**(2), 254-266.

22. Heinrich, L.J. (2002). *Informationsmanagement.* (7th ed.). Munich: Oldenbourg.

23. Henderson, M.M. (1987). The importance of data administration in information management. *Information Management Review,* **2**(4), 41-47.

24. Hoven, J. van den, (1995). Information resources management: an enterprise-wide view of data. *Information Systems Management,* **12**(3), 69-72.

25. Hoven, J. van den (2001). Information resource management: foundation for knowledge management. *Information Systems Management,* **18**(2), 80-83.

26. Huebner, H. (1996). *Informationsmanagement und strategische Unternehmensfuehrung.* Munich: Oldenbourg.

27. Ives, B. & Learmonth, G.P. (1984). The information system as a competitive weapon. *Communications of the ACM,* **27**(12), 1193-1201.

28. Kaplan, R. & Norton, D. (1992). The balanced scorecard: measures that drive performance. *Harvard Business Review,* **70**(1), 71-79.

29. Kettinger, W., Grover, V., Guha, S. & Segars, A.H. (1994). Strategic information systems revisited: a study in sustainability and performance, *MIS Quarterly,* **18**(1), 31-58.

30. Kind, J. (1986). Besseres Informationsmanagement durch externe Datenbanken. *Office Management*, (5), 490-492.

31. Kmuche, W. (1997). Informationsmanagement: chancenreiches und zukunftsrelevantes Taetigkeitsfeld mit neuen Anforderungen an die Ausbildung. *Nachrichten für Dokumentation*, **48**(3), 151-157.

32. Kroll, H. (1990). *Informationsvermittlung in der Industrie*. Cologne: RKW-Verlag.

33. Krueger, W. & Pfeiffer, P. (1988). Strategische Ausrichtung, organisatorische Gestaltung und Auswirkungen des Informationsmanagements. *Information Management*, (2), 6-15.

34. Kuhlen, R. & Finke, W. (1988). Informationsressourcen-Management: Informations- und Technologiepotentiale professionell für die Organisation verwerten. *Zeitschrift für Führung und Organisation*, **57**(5), 314-323.

35. Levitan, K.B. (1982). Information resource(s) management – information resources management. *Annual Review of Information Science and Technology*, **17**, 227-266.

36. Levitin, A.V. & Redman, T.C. (1998). Data as a resource: properties, implications, and prescriptions. *Sloan Management Review*, **40**(1), 89-101.

37. Lewis, D. & Martin, W. (1989). Information management: state of the art in the United Kingdom. *ASLIB Proceedings*, **41**(7-8), 225-250.

38. Lytle, R.H. (1986). Information resource management. *Annual Review of Information Science and Technology*, **21**, 309-336.

39. Lytle, R.H. (1988). Information resource management: a five-year perspective. *Information Management Review*, **3**(3), 9-16.

40. Marchand, D.A. & Horton, F.W. (1986). *Infotrends: profiting from your information resources*. New York, NY: Wiley.

41. Mata, F.J., Fuerst, W.L., & Barney, J. B. (1995). Information technology and sustained competitive advantage: a resource-based analysis. *MIS Quarterly*, **19**(4), 487-504.

42. Maurer, H. (2003). Wissensmanagement: ein Schritt nach vorne oder nur ein neues Schlagwort? *Informatik Spektrum*, **26**(1), 26-33.

43. McCain, K. (1990). Mapping authors in intellectual space: a technical overview. *Journal of the American Society for Information Science*, **41**(6), 433-443.

44. McFarlan, F.W. (1984). Information technology changes the way you compete. *Harvard Business Review*, **61**(3), 98-103.

45. Meik, F. (1997). Informationsmanagement reicht nicht. *Information Management*, (5), 42-47.

46. Mertens, P. & Plattfaut, E. (1986). Informationstechnik als strategische Waffe. *Information Management*, (2), 6-17.

47. Mertens, P. (1995). Wirtschaftsinformatik–von den Moden zum Trend. In Wolfgang König, (Ed.) *Wirtschaftsinformatik '95*, (pp. 25-64) Heidelberg: Physica.

48. Miller, B. B. (1988). Managing information as a resource. In J. Rabin & E.M. Jackowski (Eds.), *Handbook of information resource management*, (pp. 3-33). New York, NY: Dekker.

49. Miron, M., Cecil, J., Bradicich, K. & Hall, G. (1988). The myths and realities of competitive advantage. *Datamation*, **34**(19), 71-82.

50. Morrow, N.M. (2001). Knowledge management: an introduction. *Annual Review of Information Science and Technology*, **35**, 381-422.

51. Nonaka, I. & Takeuchi, H. (1995). *The knowledge-creating company: how Japanese companies create the dynamics of innovation*. New York, NY: Oxford University Press.

52. Noot, T.J. (1998, October). Libraries, records management, data processing: an information handling field. *ARMA Records Management Quarterly*, 22-26.

53. Otremba, G. (1987). Taetigkeitsfelder des Informationsmanagements. *Nachrichten für Dokumentation*, **38**(4), 201-203.

54. O'Brien, J.A. & Morgan, J.N. (1991). A multidimensional model of information resource management. *Information Resources Management Journal*, **4**(2), 2-11.

55. Pietsch, T., Martiny, L. & Klotz, M. (1998). *Strategisches Informationsmanagement.* (3rd ed.). Berlin: Erich Schmidt.

56. Ponzi, L. (2002): *The evolution and intellectual development of knowledge management.* Unpublished PhD dissertation. Long Island University, Palmer School of Library and Information Science.

57. Ponzi, L. & Koenig, M. (2002). Knowledge management: another management fad? *Information Research*, **8**(1). Retrieved 9 March, 2005 from http://InformationR.net/ir/8-1/paper145.html.

58. Porter, M. & Millar, V.E. (1985). How information gives you competitive advantage. *Harvard Business Review*, **63**(4), 149-160.

59. Robek, M.F., Brown, M.F. & Stephens, D.O. (1996). *Information and records management.* New York, NY: McGraw Hill.

60. Rockart, J.F. & Crescenzi, A.D. (1984). Engaging top-management in information technology. *Sloan Management Review*, **25**(4), 3-16.

61. Ross, J.W., Beath, C.M. & Goodhue, D.L. (1996). Develop long-term competitiveness through IT assets. *Sloan Management Review*, **38**(1), 31-42.

62. Savic, D. (1992). Evolution of information resource management. *Journal of Librarianship and Information Science*, **24**(3), 127-138.

63. Scheer, A-W. (1992). *Architecture of integrated information systems: foundations of enterprise modelling.* Berlin: Springer.

64. Scheid, E.M. (2001). *Wissensmanagement - mehr als eine moderne Philosophie!* Berlin: Gora, Hecken & Partner. Retrieved 9 March, 2005 from http://www.c-o-k.de/cp_artikel.htm?artikel_id=24

65. Schlögl, C. (2000). Informationskompetenz am Beispiel einer szientometrischen Untersuchung zum Informationsmanagement. In *Proceedings des Internationalen Symposiums für Informationswissenschaft* (pp. 89-111) Konstanz: Universitaetsverlag Konstanz.

66. Schlögl, C. (2003). Wissenschaftslandkarte Informationsmanagement. *Wirtschaftsinformatik*, **45**(1), 7-16.

67. Schneider, U. (1990). *Kulturbewusstes Informationsmanagement.* Munich: Oldenbourg.

68. Schneyman, A.H. (1985). Organizing information resources. *Information Management Review* , **1**(1), 35-45.

69. Schulte, U. (1987). Praktikable Ansatzpunkte zur Realisierung von Datenmanagement-Konzepten. *Information Management,* (4), 26-31.

70. Schwarze, J. (1988). Zum Berufsbild des Informations-Managers. *Information Management,* (1), 48-53.

71. Sutherland, E. (1991). Methodologies and models for information management. *ASLIB Proceedings*, **43**(2/3), 99-107.

72. Senge, P. (1990). *The fifth discipline: the art & practice of the learning organization*. New York, NY: Doubleday.

73. Stewart, T. (1997). *Intellectual capital: the new wealth of organizations.* New York, NY: Doubleday.

74. Sveiby, K. (1997). *The new organizational wealth: managing and measuring knowledge-based assets.* San Francisco, CA: Berrett Koehler.

75. Synnott, W.R. (1987a). *The information weapon: winning customers and markets with technology*. New York, NY: Wiley.

76. Synnott, W.R. (1987b). The emerging chief information officer. *Information Management Review,* **3**(1), 21-35.

77. Taylor, R.S. (1986). *Value-added processes in information systems.* Norwood, NJ: Ablex.

78. Trauth, E.M. (1989). The evolution of information resource management. *Information and Management,* **16**(5), 257-268.

79. Vickers, P. (1985). A holistic approach to the management of information. *ASLIB Proceedings* , **37**(1), 19-30.

80. Vogel, E. (1992). Informationsmanagement: Stand und Perspektiven des Managements von Informationsressourcen. In M. Buder, W. Rehfeld & T. Seeger (Eds.), Grundlagen der praktischen Information und Dokumentation (pp. 897-927). >Vol. 2. (3rd ed.). Munich: Saur.

81. Wang, R.Y. (1998). Manage your information as a product. *Sloan Management Review,* **39**(4), 95-106.

82. Wersig, G. (1989). *Organisations-Kommunikation. Die Kunst, ein Chaos zu organisieren* Baden-Baden: FBO-Fachverlag.

83. Wiggins, R.E. (1988). A conceptual framework for information resources management. *International Journal of Information Management,* **8**(1), 5-11.

84. Wilson, T.D. (2002). Information management. In J. Feather and P. Sturges, (Eds.), *International encyclopedia of information and library science,* (2nd ed.) (pp. 263-278) London: Routledge, 2002

85. Wilson, T.D. (2002). The nonsense of 'knowledge management'. *Information Research,* **8**(1). Retrieved 9 March, 2005 from http:// InformationR.net/ir/8-1/paper144.html.

86. Wiseman, C. (1985). *Strategy and computers: information systems as competitive weapons.* Homewood, IL.: Dow Jones-Irwin.

87. Wollnik, M. (1988). Ein Referenzmodell des Informations-managements. *Information Management,* (3), 34-43.

88. Zijlker, A.W. (1988). Setting the scene for I.R.M. *Information and Management,* **15**(2), 79-84.

10

INFORMATION AND KNOWLEDGE: AN EVOLUTIONARY FRAMEWORK FOR INFORMATION SCIENCE

Marcia J. Bates

Department of Information Studies, University of California, Los Angeles, Los Angeles, CA 90095-1520, USA

ABSTRACT

Background

Many definitions of information, knowledge, and data have been suggested throughout the history of information science. In this article, the objective is to provide definitions that are usable for the physical, biological, and social meanings of the terms, covering the various senses important to our field.

Argument

Information 1 is defined as the pattern of organization of matter and energy. Information 2 is defined as some pattern of organization of matter and energy that has been given meaning by a living being. Knowledge is defined as information given meaning and integrated with other contents of understanding.

Elaboration

The approach is rooted in an evolutionary framework; that is, modes of information perception, processing, transmission, and storage are seen to have developed as a part of the general evolution of members of the animal kingdom. Brains are expensive for animals to support; consequently, efficient storage, including, particularly, storage at emergent levels-for example, storing the concept of chair, rather than specific memories of all chairs ever seen, is powerful and effective for animals.

Conclusion

Thus, rather than being reductionist, the approach taken demonstrates the fundamentally emergent nature of most of what higher animals and human beings, in particular, experience as information.

PART 1: DEVELOPMENT OF BASIC CONCEPTS AND FRAMEWORK

1. INTRODUCTION

The intent of this essay is to define and conceptualize information for the purposes and uses of information science/studies. The emphasis is three-fold: 1) the information definition is presented and contextualized within an evolutionary framework, 2) *knowledge* and *data* are defined and discussed in relation to the information concept and 3) various theoretical issues surrounding this understanding of information are developed in detail. A companion essay (Bates, in press) presents the same definition, then develops and justifies a series of fundamental forms of information.

1.1 Background

A conceptualization of information is obviously central to a discipline named information science. Many major and minor efforts have been made over the years to develop the term and to provide a framework for theory development and further general development of information science (Belkin 1978; Belkin & Robertson 1976; Brier 1998; Brookes 1975, 1980; Buckland 1991; Budd 2001; Day 2001; Derr 1985; Dervin 1977, 1983, 1999; Dretske 1981; Floridi 2002;Fox 1983; Goonatilake 1991; Hjørland 2002b; Losee 1990, 1997; MacKay 1969; Meadow & Yuan 1997; Pratt 1977; Raber 2003; Shannon & Weaver 1975; Thompson 1968; among others). Major collective efforts, with many contributors, have been published in Machlup & Mansfield (1983)and Vakkari & Cronin (1992). There have been many reviews of the literature on the concept also (Aspray 1985; Fischer 1993; Wellisch 1972; Wersig & Neveling 1975; among others), of which Cornelius (2002) and Capurro & Hjørland (2003) appear to be the most recent.

A proper review of these many definitions and discussions of information would be book-length and would make impossible an adequate discussion of the definition proposed in this essay. A brief review of common classes of definitions will be presented and discussed below, while the reader is referred to other sources for a detailed discussion of other views

One source of the variety of approaches may be the many different disciplinary origins of writers on information. Engineering, the natural sciences, a wide array of the social sciences and the humanities have all contributed to the discussion. Ultimately, however, our discipline must surely find a way of thinking about information that is distinctively suitable for our own theoretical and practical uses (Bates, 1987, 1999).

Within information science, it is desirable to develop an understanding of information that is applicable for the various senses in which researchers and practitioners need to talk about information. Buckland (1991) has said that most definitions in the field represent approaches toward information that take information 1) as knowledge, 2) as process, or 3) as thing. Derr (1985), Dretske (1981) and Fox (1983) have followed the knowledge approach, by taking information to be various types of propositional statements. Boulding (1961), Brookes (1980) and Dervin (1977) have taken a different tack within the knowledge approach, by writing of information as constituting additions or changes in a mental map. Finally, after the original submission of this essay, but before the final version was sent in, a paper by A.D. Madden appeared, titled 'Evolution and information', where information is defined as 'a stimulus which expands or amends the World View of the informed' (Madden 2004: 9). This approach constitutes another of the definitions of the additions-or-changes-in-the-mental-map variety. Madden's paper is a reflection on why and how animals need to perceive and communicate information to each other.

Continuing with Buckland's categories of information definitions, Pratt (1977) has taken the purest process approach by defining information as the moment of being informed, of 'in-formation' in the mind. Buckland himself (1991) has emphasized information's *thing-ness*, always of importance in a world with many information objects lying around.

Claude Shannon, the progenitor of the original burst of interest in the concept just after World War II, took the amount of information to be a function of one's freedom of choice in selecting a message (Shannon & Weaver 1975). Cornelius (2002), Day (2001), Hayles (1999) and Rayward (1992), on the other hand, view information as a social construct and endeavor to extract out the social meanings and history surrounding the term. In order not to exhaust the space constraints of this essay, the reader is directed to the cited resources for more detail on the various conceptualizations of information.

Though a historical approach is not being taken in this essay, it is appropriate to provide some historical context to the discussion that follows, for reasons that will become apparent. If we examine the above range of types of definitions of information, we see that they can be fit within the broader pendulum swings of the sciences and humanities over the past fifty years. In our field and many others in the 1950s forward, an extreme scientism—including the most extreme form, logical positivism—dominated. Social science disciplines gained legitimacy by showing how their subject matters could be studied using adaptations of classic, natural scientific methods.

In more recent years, there has been a reaction to this approach, with a concomitant swing towards the use of what are essentially humanities methods in the social sciences. Now the fashion is to deride the very scientific techniques so recently valorized and to insist that only highly qualitative and subjectivist methods produce credible results. Hermeneutic interpretation, detailed participant observation and historical analysis, among others, are now the methods of choice. Nowadays, it is seldom remembered, however, that the logical positivist approach was itself a reaction to what were deemed ineffective subjectivist research philosophies that preceded it.

In like fashion, attitudes toward information itself have swung between highly objectivist and subjectivist interpretations. When I began teaching in the 1970s, most students in my classes would insist on the objective nature of information; now that position is reversed and most feel that information has no meaning except as it has impact on an experiencing being. Their reactions paralleled the formal literature—and perhaps the spirit of the times as well. I believe that we are missing the most important lesson that should be coming out of these historical swings—the recognition that each of these positions has something to teach us and that the long-term goal should be to develop an approach that allows each perspective to give over to us what it has to teach. Perhaps we can find a way to think about information that effectively allows for both subjective and objective perspectives.

1.2 Objectives of this essay

Currently, there are a number of metatheoretical positions, that is, fundamental philosophical and theoretical frames, or 'orienting strategies' (Wagner & Berger1985), competing for the attention of the discipline of information studies. Ellis (1992), Frohmann (1994) and Hjørland (2002a), among others, have contributed to the elucidation of these positions. They include research metatheories that are historical, statistical and bibliometric, engineering, and design-based, as well as ones arising out of sociology, psychology, and the physical sciences.

These various metatheoretical positions have contributed enormously to information studies, but there is one very productive metatheory in the social and biological sciences which has been largely ignored in information studies, and which will form the basis for the understanding of information presented herein. Over the last several decades, researchers in biology, anthropology, and psychology have come together to study human behavior in a new way. That new framework is rooted in an understanding of evolution and its impact on the cognitive, linguistic, and social structures of human beings. There are several variations on this approach, going by names such as human behavioral ecology (Winterhalder & Smith 2000), and evolutionary culture theory (Durham 1991). Probably the most common name used is evolutionary psychology (Barkow, *et al.* 1992).

These researchers have been working to integrate their disciplines' understandings of human beings in ways that promote a richer comprehension than any one perspective provides. This 'conceptual integration' (Cosmides, *et al.* 1992) in the human sciences is intended to relate and connect the knowledge among these disciplines, not reduce one field to another. No claims are being made that psychology is *merely* biology. On the contrary, this metatheory not only recognizes but also explicitly draws on the unique learning contributed by each of the biological and social sciences.

For our purposes, it is not necessary to subscribe to the full theoretical programme of evolutionary psychology (or any of its variants) to reflect on and use the ideas to be presented in this paper. However, putting these ideas in the evolutionary context enriches one's understanding and provides a possible foundation for building a scientifically-based conception of information that also harmonizes with several of the more social-science-based theories of information in our field. The ultimate goal is to find a basis for an integrative understanding of information for information science.

1.3 A definition of information

Where concepts or ideas are drawn from others' work, they are so cited; otherwise the following conceptualization is my own.

The basic definition.

An early think-piece by Edwin Parker in a conference on information needs provided the following definition of information, which is the one used in this essay.

1.4 Information is the pattern of organization of matter and energy. (Parker 1974: 10)

The definition is not original with Parker; this approach to information was endemic in the 1970s. However, in the 1970s and early 1980s, when I became interested in this definition, I was not able to find any formal description or development of it—nor any since—just occasional references to the idea. I believe that we in information science have not developed the potential of this productive definition for the field. Now that the biological and human sciences are developing a deeper and richer understanding of information processing and adaptation, there is a renewed value in considering this definition.

1.5 The definition of pattern.

Pattern is understood to be used in the above definition in two of the senses commonly found in dictionaries. Here are two of those in the Random House Unabridged Dictionary (1993: 1423). First: 'a natural or chance marking, configuration, or design: *patterns of frost on the window*'. Secondly, 'a combination of qualities, acts, tendencies, etc., forming a consistent or characteristic arrangement: *the behavior patterns of teenagers*'. The first definition implies a first-order pattern. The design we see on the frosted window is not part of any larger whole; it is simply a grouping or spotting of frost on the window that is not totally chaotic—so it does contain information—but nor is it a system. (To be very precise, ice crystals form in a regular pattern, but since frost must form on something, the irregularity of frost patterns reflects the irregularities of the surface it forms upon [Frost. Encyclopaedia Britannica online.)

The second definition implies a second-order kind of pattern, one in which a variety of features are knitted together in an overall system or integral design. A pattern of behavior implies repeated similar cycles of activity, that is, some coherence above and beyond the bunching and scattering of the frost on the window.

These patterns may be characterized as *emergent*, meaning that the sum of the elements constitutes something new, a whole with its own distinct qualities. When we look at a chair, we see a (first-order) pattern of light and dark, solid colors and edges. But we also are capable of recognizing the chair as a chair, an emergent pattern that we can recognize quickly as a whole, because it possesses certain features we have learned to recognize through experience and with the help of our inherent cognitive abilities.

This is not to imply that a sharp distinction between first- and second-order patterns is being proposed. Patterns form, dissolve, fragment, etc. in many ways continually. Rather, both definitions are provided in order to claim both senses for *pattern* in the above definition. A *pattern of organization*, as used here, does not have to imply coherence, though it often does; just something other than pure entropy.

1.6 Whose pattern of organization?

A question immediately arises upon hearing of such a definition of information. *Whose* pattern of organization is *the* pattern? Is there one objective pattern, or numerous individual subjective patterns, that is being referred to?

To proceed further with elaborating the definition, it is necessary to state two fundamental assumptions that ground the ideas that follow. The first assumption is that there is a real universe out there, that we are not solipsists, that is, what we see as being outside ourselves is not a movie we are running in our minds, but an actual universe independent of ourselves. To posit an independent universe, however, does not necessitate the further assumption that we humans have a complete, clear, or 'true' understanding of that independent universe, or that it is ultimately understandable in only one way, just that it exists in some form. The second assumption is that the universe is not total entropy, or total undifferentiated chaos or disorder. To put the latter in the affirmative, the assumption is that there are differences, distinctions, differentiations existing in that universe. For example, there are some elements or compounds in one place and other elements or compounds in a different place, thus making some parts of the universe different from other parts. In our daily lives, of course and in science and other learning, we believe that there are innumerable such differentiations.

But having posited that there is some structure in the universe independent of the experience of living creatures does not necessarily entail the assumption that there is one true structure to that universe and every variant understanding is false. Animals differ tremendously in their perceptions and in their cognitive equipment. The colour vision of cats is not as good as that of human beings, but cats have better night vision. Consequently, the world looks significantly different to them from what it looks to human beings—but it is not the case that we see the world *truly* and they see the world *falsely*. Animals react to and interpret their worlds dramatically differently from one another, depending on their inborn equipment and life experiences.

When information is defined here as the pattern of organization of matter and energy, there are patterns of organization that exist in the universe whether or not life exists anywhere in it. There is one shape and structure of a rock here and a different shape and structure of another rock there, whether or not any animals ever see the rocks. At the same time, once life comes along, it is useful for those living things to perceive and interact with their environments. How each living thing experiences its environment will have enormous variations and some similarities. My pattern of organization is not your pattern of organization, but, at the same time, we both live in the same world and may be responding to virtually the same things. The point here is that there are many patterns of organization of matter and energy; something going on in the universe independent of experiencing beings, as well as all the various perceived and experienced patterns of organization that animals develop out of their interactions with the world. All of these patterns of organization can be looked upon from an observer's standpoint as information; whether they are independent of sensing animals or are the tangible neural-pattern results of processing in an individual animal's nervous system. *In this particular sense*, both of what are usually called objective and subjective senses of pattern of organization are included in the definition as used here.

1.7 The evolutionary context

All these differentiations, clusterings and bunchings of matter and energy present in the universe can be the subject of animals' perceptions and interactions with the world, whether with animals, plants, rocks, i.e., matter, or with waves of sound or waves of the earth in an earthquake, i.e., energy. *Animals are able to detect and process those differentiations in the universe and experience them as patterns of organization.* (Plants can respond to light and other environmental phenomena too, but for simplicity's sake, only animals, with special emphasis on human beings, will be discussed.)

In information terms, an information-rich world is one in which there are many variations from pure chaos or entropy, i.e., where there are many regularities, differentiations, bunchings and clusterings of matter and energy. To be more specific, let us take as an example a typical chair in a university classroom. The chair is made of molded blue plastic with chrome metal legs. The physical materials comprising the plastic chair have a different pattern of organization from that of the air around it. The chair and the air consist of different physical elements and so reflect light differently from each other. We happen to be animals that have developed a perceptual apparatus to detect those light differences, which differences handily correspond to physical differences in matter that are likely to be of importance to us as physical organisms moving about the world. Being able to discern the chair through sight, before bumping into it and getting a bruise, enables us to be more successful animals in the world than otherwise.

These perceptual apparatuses differ enormously from species to species in their character, complexity and power. The amoeba can perceive light, the bat can bounce signals off of the walls of a cave, the hound can smell a scent days after the source of the scent has moved on and the human can read a freeway sign from fifty metres away. Through the evolutionary processes of natural selection, we develop these senses to take advantage of the differentiations around us, which, in turn, promotes the survival of ourselves and our offspring. Animals in very deep ocean, where there is no light at all, lose, or never develop, light-sensing capabilities.

We humans are even so clever that we have found a number of ways to enhance our perceptual apparatus. We do it through direct magnification, as with a hearing aid or telescope, or through more indirect means, whereby we find ways of representing patterns of organization that would otherwise be undetectable by us. For example, the astronomer develops an apparatus to detect radio waves coming from distant stars. A human being cannot detect those waves directly even when they are magnified. So we find ways to convert the features of those waves into something we can detect and decipher, such as coded numbers on a screen.

Patterns of organization are not limited to perceptions, however. In our brains we create and store our own patterns of thought, feeling and memory in the neurons, which we then subsequently draw on for further thought and action. Further, we mould the world around us by imposing patterns of organization on the world, whether by intentionally producing houses, tools, books and the like, or unintentionally by beating a path through the woods by repeated use. In our own species, we generate enormously complex social systems and cultural expressions, which are, in turn, stored as information in our minds and in the patterns of organization of the objects and infrastructure around us.

Over the billions of years of life on this planet, it has been evolutionarily advantageous for living organisms to be able to discern distinctions and patterns in their environment and then interact knowingly with that environment, based on the patterns perceived and formed. In the process of natural selection, those animals survive that are able to feed and reproduce successfully to the next generation. Being able to sense prey or predators and to develop strategies that protect one and promote the life success of one's offspring, these capabilities rest on a variety of forms of pattern detection, creation and storage. Consequently, organisms, particularly the higher animals, develop large brains and the skills to discern, cognitively process and operationally exploit information in the daily stream of matter and energy in which they find themselves.

Why should animals experience *patterns* of organization? Why not just process each bit of information, whether a pulse along a neuron in the brain, or a pixel of light on a computer screen, as itself, as just one bit? The answer is that detecting a pattern can be vastly more efficient in processing and storage than detecting and storing all the individual bits of information that comprise the pattern.

Storage efficiency is not a trivial matter; brains use a huge part of the energy an animal takes in as nourishment. As Allman (2000: 175) notes, 'in a newborn human the brain absorbs nearly two thirds of all the metabolic energy used by the entire body'. Even in human adults, the brain uses about 20 percent of the body's metabolic energy (Elia 1992: 63).

Allman also states:

Brains are informed by the senses about the presence of resources and hazards; they evaluate and store this input and generate adaptive responses executed by the muscles. Thus, when the required resources are rare, when the distribution of these resources is highly variable, when the organism has high energy requirements that must be continuously sustained and when the organism must survive for a long period of time to reproduce, brains are usually large and complex. In the broadest sense then, brains are buffers against environmental variability (2000: 2-3).

Calvin (2002) has argued that we human beings developed our large brains and capacity for generativity, planning and cooperation because of the whipsaw effect of being subjected to hundreds of cooling and warming periods during the Ice Ages. Substantial climatic changes could occur in as little as one to ten years; consequently, gradual evolution of genetic adaptation was too slow for survival. Our species may have survived because, with the development of larger and larger brains, we were able to adapt *in real time* to drastic environmental shifts. Stanley (1996), another respected scientist, makes a similar argument.

The kind and nature of patterns that we can recognize vary from species to species, and, to some degree, from individual to individual within a species. These differences have arisen in response to evolutionary pressures, including the character of ecological niches that species inhabit. See, for example, the description by Korpimäki, *et al.* (1996) of the different responses of voles (small, mouse-like rodents) to weasel predation vs. kestrel predation.

Human beings, with our large brains and extensive processing and storage capacity, can process patterns of great sophistication. Let us consider a human example. Consider the familiar concept of *bait and switch* in retail sales. In this pattern of behavior, a retailer offers an inexpensive product for sale through advertising, then when the prospective customer arrives at the store, the retailer says that the cheap item has been sold, but has some more expensive items for the customer to consider. The customer has been baited with an attractive product, then the product switched for actual sale.

Now, the person unfamiliar with bait and switch may be fooled once or twice, until she goes from processing the individual experiences to drawing the broader, categorizing conclusion that there is a *pattern across the instances.* She has now moved up from a simple collecting of retail experiences to seeing a broader, more sophisticated, emergent pattern in the interactions. From then on, she can think of a certain type of interaction with retailers by this common phrase and concept of *bait and switch.* There is a tremendous coding efficiency in this ability to see the broader patterns and she is also better protected against deceit because she can more quickly make a pattern match with her new understanding and avoid being cheated.

It is hard to imagine the use of language without the use of these categorizing capabilities. If every chair, being a different physical object from every other chair, had to be separately studied, remembered and named, in order for people to talk about it, we would quickly be bogged down in horrendous mental processing and storage overloads. Furthermore, we would not be able to call upon past experience when we encounter a new chair, because we could not group it with known objects with common features. Our ability to find useful commonalties between objects, which we then use to group related objects together and treat similarly under the same mental concept and same linguistic term, produces massive savings in cognitive processing and storage needs.

When an information definition such is this one is proposed, that is, one that starts with rudimentary physical and biological elements, a common reaction is to reject the definition on grounds that it is reductive, that is, that the purpose of the definition is to reduce processes at higher levels, such as the social and aesthetic, to their mere physical components. But as the argument above demonstrates, the *approach taken here is not only not reductive, it is actively constructive and emergent.* It is assumed that animals and human beings, particularly, have the ability to construct and remember global patterns perceptually, cognitively, kinesthetically, emotionally, etc. Rather than being reductive, this view of information is *productive*, because it assumes that processes at lower levels make possible the development of more and more sophisticated emergent patterns at higher levels, that is, patterns that are more than the sum of their parts, which we can then use for further development and understanding. The long-term history of our species has been one of understanding patterns of greater and greater levels of sophistication, which, in turn, produces ever-faster cycles of development and understanding.

Some readers of this essay in manuscript have argued that this definition of information simply equates *information* with *pattern* and that the definition is therefore trivial. I do not follow the logic: even if I were doing that, if I could make a good case for that definition and its usefulness to the field, would that not be a step forward?

In any case, my use of *pattern of organization of matter and energy* as the definition does more than define information as simply pattern. Instead, the creation in mind and actions, in perceptions and interactions, of patterns of organization at *many emergent levels* indicates the power of animal cognition and behaviour. By seeing a wide range of different objects, each with somewhat similar, but by no means identical characteristics as all being chairs, for example, we humans are able to engage in powerful and efficient cognition and behaviour. By seeing a similar but by no means identical sequence of actions as *bait and switch*, we are able to cope more effectively with our world and protect ourselves. Information is thus not just the literal atom-by-atom pattern of matter and energy; rather we can study information as existing at emergent levels as well. Surely, it enriches our understanding in information science to begin with this basic understanding of information, then see how this elemental definition contributes to ever-more-sophisticated understandings of patterns of organization, of information, in animal and human lives.

So, however emergence happens at a physical and biochemical level in mental processing, we accept that it does happen. Further, whatever else is going on in a human being's neural circuits, from the standpoint of information science, these mental abilities to detect emergent ideas and objects are immediately understandable in terms of coding efficiency and as effective and powerful means of representation. We in information science are accustomed to creating digital representations and compressing them for efficient storage and transmission through information technology; we should not be surprised that nature got there before us in making efficient use of our limited neurons.

In sum, animals evolve to be able to detect, process and store patterns of organization in their environment and experience. Such capabilities protect them against predators and enable them to find food and mates. Brain processing power is expensive for animals to support, however; consequently brain size has generally grown under conditions of extreme and rapid change or variability, where the high costs of a large brain have been repaid in offspring survival. The large brains of humans permit the detection, processing and storage of patterns of organization of great sophistication, with many emergent qualities. An understanding of information as the pattern of organization of matter and energy thus makes sense in the context of the long-term evolutionary development of human beings.

1.8 Revisiting the definition

Now that we have seen the ways in which this definition fits within an evolutionary context, let us revisit the definition and consider its role in information science.

First, why should we in information science want to place information in an evolutionary context? After all, after World War II, Shannon's work (Shannon & Weaver 1975) led many to try to conceptualize information in a physical way. Yet that application has never carried over well into the social and psychological senses in which we like to think about information. Will we not have the same problems with a biologically-based sense of information?

In response, Shannon's view of information greatly deepened and enriched our understanding of information. Space does not permit me to make the argument at length, but much of computer science and technical communication, as well as the information processing model of human cognition drew on Shannon's work. (See Gardner 1985; Lachman, *et al.* 1979; and Miller, 1951.) So the route science took through Shannon's information theory, while not ultimately wholly satisfying, was immensely enriching to our understanding of communication, information processing, and information transfer at the time. Indeed, the much deeper understanding provided by Shannon's work of how coding and combinatorial possibilities function in communication very probably made much easier the understanding of how coding works in DNA.

Similarly, I believe that the pass through evolutionary theory will greatly enrich our understanding of information as well. If we start with a physical sense that nature is not in a state of total entropy, but instead contains many differentiations of various kinds, then, we can see that once life began, it became advantageous for living things that move, particularly animals, to have a sensory capability to detect and construct patterns out of the differentiations existing in nature. In other words, if we accept that that there is some world out there, outside of animals and that animals' survival is promoted, in ways well understand in evolutionary biology, when they can detect and process inputs about that exterior world, then we can expect such things as perceptual apparatuses, nervous systems and brains to develop. Such a conception holds out the possibility for information science of an understanding of information fully as broad as the world of things that people and other animals might experience as informative.

Thus, I would argue that this conception of information marks the next step up from the earlier understanding of information developed in the era of physical information theory. A purely physical information processing model of information for human beings fails to take into account the immense number of short-cuts, storage efficiencies and processing efficiencies that have evolved over the millennia in animals.

Evolutionary psychologists argue that the characteristics of modern humans evolved and were honed during the hundreds of thousands of years that we and our predecessor hominin species lived as hunter-gatherers (Barkow, *et al.* 1992). These characteristics include brain organization and information processing. Thus, this evolutionary understanding is arguably a necessary, though not necessarily sufficient, part of a more general understanding of information for the purposes of information science.

In the next sections, we shall find that when we begin with the physical and biological definition provided herein, we are able to think about information in ways that are novel, productive and surprisingly relevant for information science. There may well be still more layers above the physical and biological levels that need to be developed for a full theory of information for the field, but we shall find that the understanding provided herein penetrates much further into the socially and psychologically important aspects of information than we might have imagined at the beginning.

1.9 So what is knowledge?

The principal intent of this essay is to define information. However, questions about the relationship of information to knowledge and data inevitably arise in the context of such a discussion. For that reason, knowledge and data are considered here and in the next section, relatively briefly, in order to demonstrate the relationship they are understood to have with information.

In the definition of information given above, information is seen to have no inherent meaning. The pattern of organization of matter and energy is just that; no more, no less. In living systems, however, things are always more complicated. Hundreds of millions of years of evolution have laid down structures associated with survival in animal brains that, in effect, give meaning to a stimulus even as the animal perceives the stimulus.

How might meaning have evolved? Mutations and other sources of genetic variation will have provided the raw material for natural selection. The animal that happens to have a more capable brain in information processing and, therefore, is able better to detect patterns of the sort discussed in this essay, will be more likely to survive and reproduce.

Earlier sections discussed how valuable it is for animals to be able to sense predators, or, using a human example, for a person to be able to detect a bait-and-switch pattern in the behaviour of another person in a sales situation. The amount and type of meaning that the vole can assign to the shadow of a predator bird overhead is presumably limited compared to what humans can ascribe. We can recognize an eagle or a hawk; to the vole, the shadow may not even be understood as coming from a bird. Rather, the shadow may simply be processed as *mortal danger*.

When I say that an animal 'assigns meaning', I mean it in this limited sense; it is in no way a conscious act of labelling, as it can be for humans. In most animals meaning assignment has developed to promote the survival of the animal, and many of its responses (including many of our own, such as 'fight or flight'), have become instinctual and automatic. The findings of evolutionary psychology suggest that more of human behaviour than we have generally been willing to acknowledge can be traced to fundamental biological survival needs and trade-offs (see Wright 1994, for an excellent introduction).

However, in higher animals and above all in human beings, there is far greater flexibility in meaning assignment than in many other animals. The squirrel, for example, can recognize the food value of the scraps in a left-over lunch sack, while other creatures may be unable to see the value in any food that comes in an unconventional form.

Much of our human brain's huge general-purpose capacity, including our unique language capacity, can be devoted to a wide range of miscellaneous contents associated with our life knowledge, our cultural and social knowledge, our family experiences, our memories and so on. A great deal of human culture and meaning assignment varies by culture or group and can take on an extraordinary complexity and variety, though rooted in fairly stable cognitive equipment. *Bait and switch* may be a pattern I should be able to recognize to promote my survival in a capitalist economy and it may be completely unnecessary and meaningless in another kind of culture.

Furthermore, over the millennia, through natural selection, life-promoting behaviour has been reinforced in various ways within surviving animals. Victory over a rival is experienced not just as the brute fact of victory. That victory, and the desire for more victories, may be reinforced in the animal by increases in hormones that promote aggressiveness, as a means of consolidating victories. In the case of human beings, the roles of various hormones are highly complex. (See, e.g., Bernhardt *et al.* 1998; Gray *et al.*, 2002.) These hormonal reinforcements have evolved to the point where we can now say that we feel good when we win. In other words, a subtle mixture of various biochemical influences affects our mood and reinforces reactions in ways that were life-supporting over the history of our development as a species.

Whether it is physical defeat of a rival in a fight, or the delivery of the perfect *mot juste* at a conference that gets a laugh from the audience, humans ascribe meaning to the events of their lives, meaning that ultimately rests in biological survival, though that meaning is very complex and heavily socially mediated in human beings. (For debates on meaning formation in biochemical and neurological senses, see Atlan & Cohen 1998; Barham 1996; Freeman 2000a, b;Langman & Cohn 1999 and Wills 1994.)

In general parlance, as well as in information studies, when we receive information from someone or something else ('Your package has arrived.' 'The Bruins won their game.') we consider ourselves *informed*. That is, we receive the natural pattern of organization of matter and energy that consists of the air moving with the sounds of someone's voice, or we read the pattern of organization of written words, or in some other way receive novel information. We then relate the sounds or letters to words and meaning in our mind and assign meaning to the communication.

In Bates (in press), to account for this process, two definitions of information are provided:

- Information 1: The pattern of organization of matter and energy.
- Information 2: Some pattern of organization of matter and energy given meaning by a living being.

In this way, we account for both the foundational sense of information, in which information is understood to exist throughout the universe, whether or not humans or other living beings perceive or use it, *and* the social and communicatory sense in which living beings interpret and give meaning to information. Once some interpretation has occurred, the information 2 may be integrated with the rest of the creature's neural stores or understanding. Thus, we define:

Knowledge: information given meaning and integrated with other contents of understanding. (Compare Kochen 1983, who simply defines knowledge as information given meaning.)

Less developed animals are capable of assigning less meaning to their environments, but from very low on the developmental scale, animals do assign meaning of some sort. The goat knows good moss when it sees it and knows to move toward it. Once our species developed enough to have a general-purpose and more or less conscious and deliberate meaning-assignment capability and, especially, developed language, then we could and do assign meaning to all kinds of things. So we developed religion and rituals, science and art and all the other cultural forms.

So how can we think about the relationship between information and knowledge? By the above definitions, for a batch of information 1 to become knowledge, it must first be assigned meaning by a living animal, at least, within the limit of that animal's capacity to assign meaning. This we call information 2. When the books are sitting in the closed library over the holiday and no one is reading them, they all contain information 1. Only when an understanding human (since no sub-human animal can understand books) is actually reading the information 1, does it become information 2. Its contents are assigned meaning in the act of reading. Thus, the reader is *informed* in the act of reading. Some or all of that information 2 may ultimately be integrated more or less permanently into the pre-existing knowledge stores of the animal, i.e., become, simply, knowledge.

Knowledge in inanimate objects, such as books, is really only information 1, a pattern of organization of matter and energy. When we die, our personal knowledge dies with us. When an entire civilization dies, then it may be impossible to make sense out of all the information 1 left behind; that is, to turn it into information 2 and then knowledge.

1.10 Finally, what are data?

Outside of information science, the term *data* conventionally refers to information gathered for processing or decision-making. Within the field, data are commonly seen as something raw, on a path to being fully *cooked* or distilled (Hammarberg 1981), that is, in a sequence that goes from data to information to knowledge to wisdom. There are numerous versions of this sequence; see Houston & Harmon (2002) for a recent one.

Within the context provided by this essay, we can see data in two more precise ways. First, data 1 may be seen as that portion of the entire information environment available to a sensing organism that is taken in, or processed, by that organism. What is too distant or otherwise outside the sensing purview of the animal cannot be data, it is simply information 1. Thus, strictly speaking, in the terms used here, animals perceive data, not information. Data 1 become information 2 when they are given meaning, then become knowledge when they are integrated (to at least some degree) with pre-existing knowledge in the mind or brain.

The second sense of data, or data 2, refers to information selected or generated by human beings for social purposes. Human beings, especially in the last several hundred years and, especially, using scientific methods, have developed a remarkable capacity to manipulate portions of the world in such a way as to deliberately generate additional information, which they experience as data 2, from which they learn immense amounts. Data in this sense may be information generated by as minimal an act as a child poking a grasshopper to see what it will do, all the way through many more sophisticated attempts to generate new understanding. The initial products of all of research and scholarship fall under this sense of data. These data are entirely generated through the actions of the researcher, whether it is the astronomer aiming a telescope and capturing the resulting images, or a social scientist observing a group of people and recording those observations. Though the observed stars may have been there all along, or the behaviour of the studied type may have been going on at other times as well, this information is not data (of the second type) until some human being has generated or captured it for purposes of gaining knowledge.

Data 2 may also be selected or generated for other social purposes. We create databases to store collections of facts about people, artifacts, etc. for the purposes of business, government, or other organizations. All the variants of this second sense of data have in common the fact that they are socially generated and socially meaningful.

PART 2: THEORETICAL AND PHILOSOPHICAL ISSUES

Various theoretical and philosophical issues relating to this definition of information are discussed below, as responses to a series of questions.

2. IS INFORMATION MATERIAL?

Two questions: 1) Is information an abstract entity, independent of the physical materials that compose it? In other words, does information exist on a sort of Platonic abstract plane, independent of physical reality? 2) Or, is information inextricable from the material that composes it? In other words, if a piece of granite contains a pattern of organization, would we then say that the granite *is* information?

It is argued here that neither of these characterizations is accurate. With regard to the first point above, the position taken here is fully materialist, that is, no abstract plane is assumed to house or manifest the information associated with the physical realities we experience. If the information is anywhere, it resides in the physical realities of nature, whether in the structure of a piece of granite, or in the neural pathways of the brain.

To say this, however, is not to say that information is *identical* with the physical materials or waves that make up the pattern of organization. *The information is the pattern of organization of the material, not the material itself.* As Wiener (1961: 132) has said: 'Information is information, not matter or energy'.

The ability to discern that aspect which we, as human beings, choose to label *pattern of organization* comes about because we are animals with sufficiently developed nervous systems and sensory apparatuses that we can think about the patterns of the sensory inputs we are receiving independently of their material. Thus, we can, for example, make a drawing of something we are looking at. We can, further, evaluate whether our drawing is a good representation of the item we have depicted, i.e., does the drawing constitute a reasonable mapping of the object from three to two dimensions? The information may now be manifested in three places—the physical object, in our mind and on the drawing essay. But it is still always in the physical world and we are dealing with the pattern in some independence from the material. (See Hektor 1999, for a discussion of the immateriality of information and Hayles 1999, for an insightful analysis of the consequences of separating information from physical embodiment in the intellectual fashions of our culture.)

These patterns are known to us, because we, like other organisms, have developed evolutionarily to be able to discern and construct patterns of organization and take advantage of them. We can recognize the pattern of organization of a rock, or the pattern of behaviour of a lemur we are studying in the wild, because we have developed the sensory and cognitive apparatus to have observational experiences and to articulate our experiences. The rock does not know of information at all. The lemur uses information, but does not know it is doing so. With few exceptions, only humans are aware that we are using information from an observed structure or event separately from the observed structure or event.

Stonier (1997), also working from an evolutionary perspective, defines information differently:

[I]information, like energy, is conceived of as a basic property of the universe; and like energy, which is traditionally defined operationally as

possessing the capacity to perform work, so information is defined operationally as possessing the capacity to organize a system (p. 1).

I would say that the information is the order in the system, not the capacity to create it. Information is not a force that does things to natural objects. Rather, it is the pattern of organization to be found in any matter or energy, moving or still.

3. IS THE DEFINITION TOO INCLUSIVE?

Some have said that this definition of information includes 'everything' and that the definition is, therefore, meaningless. First, it does not include everything, because it does not include total entropy, undifferentiated chaos, which, by definition, has no pattern or order. Second, information has not been defined as 'everything'. Rather, information is *the pattern of organization of everything.* As argued in the previous section, information is distinct from matter and energy (Wiener 1961: 132). The value and conceptual productivity of this more specific definition was discussed at length in the section, 'The evolutionary context', above. Third, if animals and humans process all the things that this definition of information includes and derive meaningful interpretations of them, then, surely, anything that they can so interpret should be included in a comprehensive and fundamental definition of information for information science.

Do not human beings potentially remember, respond, or act on any conceivable bit of information encountered? Hitting your knee on a rock, or hearing your friend's question, as well as countless other things and experiences can all be informative. So why should any of these information structures be eliminated in a definition of information?

What really matters for the development of information science as a discipline is whether such a definition can be successfully employed and built upon in the field's theory and practice. In effect, this entire essay and its companion essay, 'Fundamental forms of information' (Bates, in press), constitute an effort to demonstrate the potentialities within this definition.

It has also been argued by some that only patterns of organization that have actually been experienced by a person should be considered information, that is, information is a meaningful concept only in relation to subjective experience. By that line of reasoning, if some signal or data has no impact on a person somewhere sometime, then there is no information in that signal or data. But are not many things *potentially* informative? Should we eliminate those from our definition of information because they have not actually informed a person yet? Further, how can we realistically study information-seeking behaviour, if we cannot consider as at least potentially informative all those things a person has *not yet* been informed by?

4. HOW DOES THIS INFORMATION DEFINITION RELATE TO EMERGENCE?

Following Claude Shannon, who produced his ground-breaking work on means to measure the amount of information after World War II, many efforts went into adapting that understanding to more cognitively and socially-based perspectives on information. Though much was learned from those efforts, on the whole, they did not succeed in forming a basis upon which information science could build a broad understanding of information and knowledge. The approach taken in this essay may take us a step closer to that development, if not all the way.

Decades of research in evolutionary biology and psychology on human development over hundreds of thousands of years has made clear the extent to which humans are *not* simple straightforward information processing machines, processing so and so many bits of information, as a computer does, to get through life. Instead, there has been more and more recognition that human beings are mixtures of capabilities, some very highly developed and efficient, others slower and clumsy. For example, we are a very social animal and can detect extremely small distinctions in the appearance of people's faces and have a part of our brain devoted to that capability, while we cannot distinguish people nearly so well by the look of their knees or hands. The latter may be nearly as large as faces, but we do not have the mental equipment to make the key distinctions we need using those as indicators of identity.

We have evolutionary short-cuts built into our cognitive processing, so we know to flee fire, for example, without thinking practically at all. But in the case of a slower-developing crisis, one not typically a part of our hunter-gatherer inheritance, we may not take it seriously or miss the warning signs altogether until it is too late.

Information 2 has been defined as the pattern of organization of matter and energy given meaning. This initial meaning assignment, before the information is integrated with the stores of knowledge, can occur relatively instantly in cases where the association among elements is well developed. Human language capabilities are so well developed that most of the time when we look at a chair, we think and speak of it as a chair *first* and not as a pattern of edges and light and dark, though we can look at it in the latter sense too, if we want. As an experienced adult in a capitalist society, I can recognize certain sequences of behaviour and circumstances as *bait and switch* shortly after entering a store. I experience my interactions with the salesperson with this extremely compact, efficient, *emergent* characterization of *bait and switch*, not as a long sequence of vocal and behavioral interactions.

In other cases, emergent understandings may be inherently more difficult or for some other reason take longer to form. For example, in dealing with someone who has a clever, subtly manipulative personality, it may take numerous interactions over years before we grasp how they relate to us and others and conclude that we should not trust them. Many new ideas in science and the arts took a long time to emerge because, despite our great intelligence as a species, something about those ideas so violated conventional thinking, that we were unable to develop that emergent understanding. With many such ideas, however, once developed, they seem terribly obvious: 'Why didn't we think of that sooner?!'

Thus, it is argued that this understanding of information, drawing on an evolutionary psychology approach, takes information science's understanding of information from the physical up to the biological and anthropological. Further, by thinking in terms of emergence, we can see that many actions and expressions can be efficiently stored and talked about in terms of their emergent meanings, while still using the key definitions for information, data and knowledge that were presented in this essay.

5. CAN SOCIAL SCIENCE METATHEORIES IN INFORMATION STUDIES BE RECONCILED WITH THE APPROACH TAKEN HERE?

In recent years, there has been a healthy debate in information studies over what the philosophical underpinnings of the field should be (Bates 1999; Budd 2001;Dick, 1995; Frohmann, 1994; Hjørland, 1998, 2002a, b; Hjørland & Albrechtsen, 1995; Talja, *et al.* 1999; Tuominen*et al.* 2002; Wiegand, 1999). Over the history of this field, people from many different disciplinary traditions have brought their worldviews and interests to bear on information-related questions. Humanities, social science and physical science and engineering approaches all have deep roots in information studies. Though sophisticated social science research dates back to the 1930s in the field, the growing interest in studying people and their interactions with information over the last thirty years has eventually led to a very large social science presence. The immediate reaction of many people to the ideas in this essay is that those ideas are purely physical and sensory and do not have much to do with the core interests in nformation seeking and other social science questions in the field. It is to be hoped that the arguments and examples provided here and in the companion essay, (Bates, in press), have gone some way to dispel that assumption.

However, there are deeper questions lurking here as well, regarding the underlying 'metatheory', or fundamental theoretical and philosophical outlook of this essay. Recent papers by Tuominen, *et al.* (2002) and Talja, *et al.* (2005) have identified a total of four principal world views that they find to be operative in this field over the last thirty years or so. They name these the information transfer model, constructivism, collectivism and constructionism. They view these four metatheories as emphasizing, respectively, the information, the individual's cognitive viewpoint, the combined socio-cognitive viewpoint, and the social and dialogic aspects of knowledge formation.

Talja, *et al.* (2005) argue that each metatheory has distinct applications that are most useful in tackling particular kinds of questions. I would agree and would add that even though there may be conflicts and disagreements, it is worthwhile to work at integrating the research results of the applications of these metatheories and even, perhaps, ultimately, the metatheories themselves.

Cosmides, *et al.*, working from the standpoint of evolutionary psychology, argue for what they call conceptual integration in the natural and social sciences. In the context of this essay, theirs constitutes a fifth metatheoretical perspective. Specifically, they state:

> ***Conceptual integration**... [italics in the original] refers to the principle that the various disciplines within the behavioral and social sciences should make themselves mutually consistent and consistent with what is known in the natural sciences as well... The natural sciences are already mutually consistent: the laws of chemistry are compatible with the laws of physics, even though they are not reducible to them... Such is not the case in the behavioral and social sciences. Evolutionary biology, psychology, psychiatry, anthropology, sociology, history and economics largely live in inglorious isolation from one another (Cosmides et al. 1992: 4).*

The authors provide a number of examples of ways in which the behavioral and social sciences can help each other and, in some cases, prevent dead-end research paths in one discipline based on pre-suppositions that have been definitively refuted in other disciplines. They go on to argue:

> *Conceptual integration generates this powerful growth in knowledge because it allows investigators to use knowledge developed in other disciplines to solve problems in their own (Cosmides et al. 1992: 12).*

They continue:

> *At present, crossing such boundaries is often met with xenophobia, packaged in the form of such familiar accusations as 'intellectual imperialism' or 'reductionism'. But... we are neither calling for reductionism nor for the conquest and assimilation of one field by another... In fact, not only do the principles of one field not reduce to those of another, but by tracing the relationships between fields, additional principles often appear. Instead, conceptual integration simply involves learning to accept with grace the irreplaceable intellectual gifts offered by other fields (Cosmides et al. 1992: 12).*

In short, they are arguing for working toward the goal of uniting all human knowledge, and doing so by examining what each field has to offer. To date, the exciting new work that is going on in a cluster of new fields: evolutionary culture theory (Durham 1991), evolutionary psychology (Barkow, *et al.* 1992), evolutionary neuroscience (La Cerra & Bingham 2002) and human behavioral ecology (Winterhalder & Smith 2000), have been largely ignored in many of the social sciences, including within information science. (In our field, Brier 1996, Cooper

2001, Madden 2004, Sandstrom 1994, 1999, 2001 and Sandstrom & Sandstrom 1995, are exceptions, having drawn on this literature.) This essay has constituted a very preliminary and novice effort to draw on that perspective, but within an orientation that is meaningful for our field.

Some have insisted that the integrative perspective is fundamentally at odds with the several metatheories described by Tuominen, *et al.* (2002) and Talja, *et al.*(2005), that these metatheories cannot be reconciled with a scientific view. Cosmides, *et al.* (1992) would suggest that such a claim of incompatibility would speak against, rather than for, those other metatheories. They would argue that the long-term objective is surely to unite all human knowledge, not to argue that certain perspectives are so unique and special that they cannot be held in mind at the same time as perspectives and knowledge from other disciplines.

Again, having an understanding of psychology that is consonant with biology does not reduce psychology to biology, nor anthropology to psychology, nor sociology to anthropology, etc. Key findings about cultures and societies cannot be predicted from psychology, nor can psychology be predicted from biology, any more than key findings in biology can be predicted from chemistry or physics. Each discipline rests within the discipline(s) at a more fundamental level, but does not consist solely of, nor is reducible to, the discipline(s) below it.

In information studies, we came late to seeing the value in the several newer social science metatheories described by Talja, *et al.* (2005); these perspectives were well developed in other fields decades before we began to draw on their literatures. We have much yet to absorb.

But evolutionary psychology and related new initiatives have already been developing for decades as well. See references in Barkow, *et al.* (1992); Clark (1997); Damasio (1999); Durham (1991); Jackendoff (2002); Johnson (1987);Knight, *et al.* (2000); La Cerra & Bingham (2002); Sperber (1996); Winterhalder & Smith (2000). These ideas have also seen application in many fields both inside and outside the sciences. See, for example, a sampling of recent papers drawing on this perspective in literature, history, social science, and philosophy (Carroll 1999; Kenrick *et al.* 2002; Lachapelle 2000; Mallon & Stich 2000;Plotkin 2003; Siegert & Ward 2002; Stuart-Fox 1999; Sugiyama 2001; Waal 2002). Indeed, the book edited by Barkow, *et al.* (1992) and its nineteen component papers have already been cited collectively over 1600 times. Shall we fall decades behind on this evolutionary approach as well, as we did with the other social science metatheories, or do we begin to draw from it now?

6. SUMMARY AND CONCLUSIONS

6.1 Summary

Many definitions of information have been suggested throughout the history of information science. In this essay, the objective has been to provide a definition that is usable for the physical, biological and social meanings of the term, covering the various senses important to our field.

Information has been defined as the *pattern of organization of matter and energy*. Information is everywhere except where there is total entropy. Living beings process, organize and ascribe meaning to information. Some pattern of organization that has been given meaning by a living being has been defined as information 2, while the above definition is information 1, when it is desirable to make the distinction. Knowledge has been defined as information given meaning and integrated with other contents of understanding. Meaning itself is rooted ultimately in biological survival. In the human being, extensive processing space in the brain has made possible the generation of extremely rich cultural and interpersonal meaning, which imbues human interactions. (In the short term, not all meaning that humans ascribe to information is the result of evolutionary processes. Our extensive brain processing space also enables us to hold beliefs for the short term that, over the long term, may actually be harmful to survival.)

Data 1 has been defined as that portion of the entire information environment (including internal inputs) that is taken in, or processed, by an organism. Data 2 is that information that is selected or generated and used by human beings for research or other social purposes.

This definition of information is not reductive--that is, it does not imply that information is all and only the most microscopic physical manifestation of matter and energy. Information principally exists for organisms at many *emergent* levels. A human being, for example, can see this account as tiny marks on a piece of paper, as letters of the alphabet, as words of the English language, as a sequence of ideas, as a genre of publication, as a philosophical position and so on.

Thus, patterns of organization are not all equal in the life experience of animals. Some types of patterns are more important, some less so. Some parts of patterns are repetitive and can be compressed in mental storage. As mental storage space is generally limited and its maintenance costly to an animal, adaptive advantage accrues to the species that develops efficient storage. As a result, many species process elements of their environment in ways efficient and effective for their particular purposes; that is, as patterns of organization that are experienced as emergent wholes.

We see a chair as a chair, not only as a pattern of light and dark. We see a string of actions by a salesperson as *bait and switch*, not just as a sequence of actions. We understand a series of statements as parts of a whole philosophical argument, not just as a series of sentences. The understanding of information embraced here recognizes and builds on the idea that these emergent wholes are efficient for storage and effective for the life purposes of human beings as successful animals (to date) on our planet.

Thus, people experience their lives in terms of these emergent objects and relations, for the most part. Likewise, information is stored in retrieval systems in such a way that it can be represented to human beings in their preferred emergent forms, rather than in the pixels or bits in which the information is actually encoded within the information system.

6.2 Conclusions

I have long felt that to succeed in the process of developing a broadly applicable, encompassing understanding of information for our field, we must begin at the physical and biological levels and move up to the cultural, social, cognitive and aesthetic. Beginning with the physical substrates of our bodies and of our cultures in defining information does not mean stopping there. Instead, we are thereby enabled to develop a truly grounded understanding of the core concept of our discipline--an understanding that can be built on and enriched indefinitely much more by drawing on the learning from the several other metatheoretical approaches of interest in information studies.

Thus, to take specific examples from the thinking of information studies, the construction of emergent patterns of organization can be seen as the general process at the root of Dervin's (1999) concept of 'sensemaking' , in which people are seen as seeking information in order to make sense of situations in their lives. 'Making sense' consists operationally in constructing an understanding, an emergent, simplifying organization of ideas and experience that satisfies people's sense of completion, of wholeness. Likewise, the construction of such emergent patterns is also at the heart of Kuhlthau's (2004) information search process, in which the searcher gradually forms a more and more coherent conception of the focus, theme, or argument of the planned paper. The formulation of an 'Anomalous State of Knowledge' (Belkin, *et al.* 1982) is in response to a situation where the information seeker feels a lack of success in creating a coherent, emergent, understanding around some phenomenon of interest and attempts to describe the shape of the gap in knowledge.

Similarly, at the social, rather than individual level, people create coherent, social groupings utilizing emergent constructions of their situations. Drawing on Hjørland & Albrechtsen's (1995) sense of *domain* , we can imagine a group of sociologists studying, say, the intersection of class and race. They create a social and intellectual domain, a specialty, that is structured socially in various forms useful to researchers, such as in professional organizations and listservs. There is a body of agreed-upon learning, as well as a research philosophy and methodology, that is, a paradigm (Kuhn 1970), motivating and uniting the group of researchers, as well as publications, laboratories and other social institutions. This entire complex of social arrangements and information forms is a domain. We are now recognizing that to understand information seeking and use, we must examine all the forms of information that compose the emergent entity we call a domain—the social arrangements, information genres and behavioral patterns.

Finally, the discourses of the constructionist metatheory that Talja, *et al.* (2005) refer to, that is, the social and intellectual arrangements of language and language patterns used by a group of people, can be understood, in the terms used here, as one means of constructing a useful emergent whole, a world view about some part of life expressed through language.

I see no conflict between a definition of information as the pattern of organization of matter and energy and our social understanding of human beings constructing all manner of emergent patterns of importance to themselves. Whether called sense, focus, or ASK, whether called domains, paradigms, or discourses, these information structures constitute coherent, emergent wholes, built up out of the information described herein.

Much remains to be done to develop the fullness of an understanding of information for this field. This essay will, I hope, contribute to that effort.

ACKNOWLEDGEMENTS

My thanks to Suresh Bhavnani, Terry Brooks, Michael Buckland, Michele Cloonan, William Cooper, Laura Gould, Jenna Hartel, Robert M. Hayes, P. Bryan Heidorn anders Hektor, Birger Hjørland, Peter Ingwersen, Jarkko Kari, Jaana Kekäläinen, Kimmo Kettunen, Erkki Korpimäki, Poul S. Larsen, Mary N. Maack, Marianne L. Nielsen, Pamela Sandstrom, Reijo Savolainen, Dagobert Soergel, Sanna Talja, Pertti Vakkari and the anonymous reviewers for their thoughtful and insightful comments on this and the companion essay (Bates, in press).

REFERENCES

1. Allman, J.M. (2000). *Evolving brains.* New York, NY: Scientific American Library.

2. Aspray, W. (1985). The scientific conceptualization of information: a survey. *Annals of the History of Computing*, **7**(2), 117-140.

3. Atlan, H. & Cohen, I.R. (1998). Immune information, self-organization and meaning. *International Immunology*, **10**(6), 711-717.

4. Barham, J. (1996). A dynamical model of the meaning of information. *BioSystems*, **38**(2-3), 235-241.

5. Barkow, J.H., Cosmides, L. & Tooby, J. (Eds.). (1992). *The adapted mind: evolutionary psychology and the generation of culture.* New York, NY: Oxford University Press.

6. Bates, M.J. (1987). Information: the last variable. *Proceedings of the 50th ASIS Annual Meeting*, **24**, 6-10.

7. Bates, M. J. (1999). The invisible substrate of information science. *Journal of the American Society for Information Science*, **50**(12), 1043-1050.

8. Bates, M.J. (in press). Fundamental forms of information. *Journal of the American Society for Information Science and Technology.*

9. Belkin, N.J. (1978). Information concepts for information science. *Journal of Documentation*, **34**(1), 55-85.

10. Belkin, N.J., Oddy, R.N. & Brooks, H.M. (1982). ASK for information retrieval: Part 1: Background and theory. *Journal of Documentation*, **38**(2), 61-71.

11. Belkin, N.J. & Robertson, S.E. (1976). Information science and the phenomenon of information. *Journal of the American Society for Information Science*, **27**(4), 197-204.

12. Bernhardt, P.C., Dabbs, J.M., Fielden, J.A. & Lutter, C.D. (1998). Testosterone changes during vicarious experiences of winning and losing among fans at sporting events. *Physiology & Behavior*, **65**(1), 59-62.

13. Boulding, K.E. (1961). *The image: knowledge of life and society.* Ann Arbor, MI.: University of Michigan Press.

14. Brier, S. (1996). Cybersemiotics: A new interdisciplinary development applied to the problems of knowledge organisation and document retrieval in information science.*Journal of Documentation*, **52**(3), 296-344.

15. Brier, S. (1998). Cybersemiotics: a transdisciplinary framework for information studies. *Biosystems*, **46**(1-2), 185-191.

16. Brookes, B.C. (1975). The fundamental problem of information science. In V. Horsnell (Ed.), *Informatics 2: Proceedings of a*

Conference Held by the Aslib Coordinate Indexing Group (Oxford, March 25-27, 1974). (pp. 42-49). London: Aslib.

17. Brookes, B. C. (1980). The foundations of information science. Part I. Philosophical aspects. *Journal of Information Science*, **2**(3-4), 125-133.

18. Buckland, M.K. (1991). Information as thing. *Journal of the American Society for Information Science*, **42**(5), 351-360.

19. Budd, J.M. (2001). *Knowledge and knowing in library and information science: a philosophical framework.* Lanham, MD: Scarecrow Press.

20. Calvin, W.H. (2002). *A brain for all seasons: human evolution and abrupt climate change.* Chicago, IL: University of Chicago Press.

21. Capurro, R. & Hjørland, B. (2003). The concept of information. *Annual Review of Information Science and Technology*, **37**, 343-411.

22. Carroll, J. (1999). The deep structure of literary representations. *Evolution and Human Behavior*, **20**(3), 159-173.

23. Clark, A. (1997). Being there: putting brain, body and world together again. Cambridge, MA: The MIT Press.

24. Cooper, W. S. (2001). The evolution of reason: logic as a branch of biology. Cambridge: Cambridge University Press.

25. Cornelius, I. (2002). Theorizing information for information science. *Annual Review of Information Science and Technology*, **36**, 393-425.

26. Cosmides, L., Tooby, J. & Barkow, J.H. (1992). Introduction: evolutionary psychology and conceptual integration. In J.H. Barkow, L. Cosmides, & J. Tooby (Eds.), *The adapted mind: evolutionary psychology and the generation of culture.* (pp. 3-15). New York, NY: Oxford University Press.

27. Damasio, A. (1999). *The feeling of what happens: body and emotion in the making of consciousness.* San Diego, CA: Harvest.

28. Day, R.E. (2001). *The modern invention of information*. Carbondale, IL: Southern Illinois University Press.

29. Derr, R.L. (1985). The concept of information in ordinary discourse. *Information Processing & Management*, **21**(6), 489-499.

30. Dervin, B. (1977). Useful theory for librarianship: Communication, not information. *Drexel Library Quarterly*, **13**(3), 16-32.

31. Dervin, B. (1983). Information as a user construct: the relevance of perceived information needs to synthesis and interpretation. In S.A. Ward & L.J. Reed (Eds.),*Knowledge structure and use: implications for synthesis and interpretation*. (pp. 153-183). Philadelphia, PA: Temple University Press.

32. Dervin, B. (1999). On studying information seeking methodologically: the implications of connecting metatheory to method. *Information Processing and Management*,**35**(6), 727-750.

33. Dick, A.L. (1995). Library and information science as a social science: neutral and normative conceptions. *Library Quarterly*, **65**(2), 216-235.

34. Dretske, F.I. (1981). *Knowledge and the flow of information*. Cambridge, MA: The MIT Press.

35. Durham, W.H. (1991). *Coevolution: genes, culture and human diversity*. Stanford, CA: Stanford University Press.

36. Elia, M. (1992). Organ and tissue contribution to metabolic rate. In J.M. Kinney & H.N. Tucker (Eds.), *Energy metabolism: tissue determinants and cellular corollaries*.(pp. 61-79). New York, NY: Raven Press.

37. Ellis, D. (1992). The physical and cognitive paradigms in information retrieval research. *Journal of Documentation*, **48**(1), 45-64.

38. Fischer, R. (1993). From transmission of signals to self-creation of meaning: transformations in the concept of information. *Cybernetica*, **36**(3), 229-243.

39. Floridi, L. (2002). What is the philosophy of information? *Metaphilosophy*, **33**(1-2), 123-145.

40. Fox, C. J. (1983). *Information and misinformation: an investigation of the notions of information, misinformation, informing and misinforming.* Westport, CN: Greenwood Press

41. Freeman, W. J. (2000a). *How brains make up their minds.* New York, NY: Columbia University Press.

42. Freeman, W. J. (2000b). A neurobiological interpretation of semiotics: meaning, representation and information. *Information Sciences*, **124**(1-4), 93-102.

43. Frohmann, B. (1994). Discourse analysis as a research method in library and information science. *Library & Information Science Research*, **16**(2), 118-138.

44. Frost. (2005). Encyclopaedia Britannica. Retrieved 1 June, 2005, from Encyclopaedia Britannica Online. http://search.eb.com/eb/article?eu=36165 [Subscription required]

45. Gardner, H. (1985). *The mind's new science: a history of the cognitive revolution.* New York, NY: Basic Books.

46. Goonatilake, S. (1991). *The evolution of information: lineages in gene, culture and artefact.* London: Pinter.

47. Gray, P.B., Kahlenberg, S.M., Barrett, E.S., Lipson, S.F. & Ellison, P.T. (2002). Marriage and fatherhood associated with lower testosterone in males. *Evolution and Human Behavior*, **23**(3), 193-201.

48. Hammarberg, R. (1981). The cooked and the raw. *Journal of Information Science*, **3**(6), 261-267.

49. Hayles, N.K. (1999). *How we became posthuman: virtual bodies in cybernetics, literature and informatics.* Chicago, IL: University of Chicago Press.

50. Hektor, A. (1999). Immateriality: on the problem of information. *KFB - Rapport,* (24) 134-162.

51. Hjørland, B. (1998). Theory and metatheory of information science: a new interpretation. *Journal of Documentation,* **54**(5), 606-621.

52. Hjørland, B. (2002a). Epistemology and the socio-cognitive perspective in information science. *Journal of the American Society for Information Science and Technology,* **53**(4), 257-270.

53. Hjørland, B. (2002b). Principia informatica: foundational theory of information and principles of information services. In H. Bruce, R. Fidel, P. Ingwersen, & P. Vakkari (Eds.), *Emerging Frameworks and Methods: Proceedings of the Fourth International Conference on Conceptions of Library and Information Science (CoLIS4)*(pp. 109-121). Greenwood Village, CO: Libraries Unlimited.

54. Hjørland, B. & Albrechtsen, H. (1995). Domain analysis: toward a new horizon in information science, *Journal of the American Society for Information Science,* **46**(6), 400-425.

55. Houston, R.D. & Harmon, E.G. (2002). Re-envisioning the information concept: systematic definitions. In H. Bruce, R. Fidel, P. Ingwersen, & P. Vakkari (Eds.),*Emerging frameworks and methods: proceedings of the fourth International Conference on Conceptions of Library and Information Science (CoLIS4)* (pp. 305-308). Greenwood Village, CO: Libraries Unlimited.

56. Jackendoff, R. (2002). *Foundations of language: brain, meaning, grammar, evolution.* New York, NY: Oxford University Press.

57. Johnson, M. (1987). *The body in the mind: the bodily basis of meaning, imagination and reason.* Chicago, IL: University of Chicago Press.

58. Kenrick, D.T., Maner, J.K., Butner, J., Li, N.P., Becker, D.V. & Schaller, M. (2002). Dynamical evolutionary psychology: mapping the domains of the new interactionist paradigm. *Personality and Social Psychology Review*, **6**(4), 347-356.

59. Knight, C., Studdert-Kennedy, M. & Hurford, J.R. (Eds.). (2000). *The evolutionary emergence of language: social function and the origins of linguistic form*. Cambridge: Cambridge University Press.

60. Kochen, M. (1983). Library science and information science: Broad or narrow? In F. Machlup & U. Mansfield (Eds.), *The study of information: interdisciplinary messages*. (pp. 371-377). New York, NY: Wiley.

61. Korpimäki, E., Koivunen, V. & Hakkarainen, H. (1996). Microhabitat use and behavior of voles under weasel and raptor predation risk: predator facilitation?*Behavioral Ecology*, **7**(1), 30-34.

62. Kuhlthau, C.C. (2004). *Seeking meaning: a process approach to library and information services.* (2nd ed.) Westport, CN: Libraries Unlimited.

63. Kuhn, T. S. (1970). *The structure of scientific revolutions*. (2nd ed.). Chicago, IL: University of Chicago Press.

64. La Cerra, P. & Bingham, R. (2002). *The origin of minds: evolution, uniqueness and the new science of the self.* New York, NY: Harmony Books.

65. Lachapelle, J. (2000). Cultural evolution, reductionism in the social sciences and explanatory pluralism. *Philosophy of the Social Sciences,* **30**(3), 331-361.

66. Lachman, R., Lachman, J.L. & Butterfield, E.C. (1979). *Cognitive psychology and information processing: an introduction*. Hillsdale, N.J.: Lawrence Erlbaum Associates.

67. Langman, R. E. & Cohn, M. (1999). Away with words: commentary on the Atlan-Cohen essay 'Immune information, self-organization and meaning'. *International Immunology*, **11**(6), 865-870.

68. Losee, R. M., Jr. (1990). *The science of information: measurement and applications.* San Diego, CA: Academic Press.

69. Losee, R. M. (1997). A discipline independent definition of information. *Journal of the American Society for Information Science,* **48**(3), 254-269.

70. Machlup, F. & Mansfield, U. (Eds.). (1983). *The study of information: interdisciplinary messages.* New York, NY: Wiley.

71. MacKay, D. M. (1969). *Information, mechanism and meaning.* Cambridge, MA: MIT Press.

72. Madden, A.D. (2004). Evolution and information. *Journal of Documentation,* **60**(1): 9-23.

73. Mallon, R. & Stich, S. P. (2000). The odd couple: the compatibility of social construction and evolutionary psychology. *Philosophy of Science,* **67**(1), 133-154.

74. Meadow, C. T. & Yuan, W. (1997). Measuring the impact of information: defining the concepts. *Information Processing and Management,* **33**(6), 697-714.

75. Miller, G.A. (1951). *Language and communication.* New York, NY: McGraw-Hill.

76. Parker, E. B. (1974). Information and society. In C. A. Cuadra & M.J. Bates (Eds.), *Library and information service needs of the nation: proceedings of a Conference on the Needs of Occupational, Ethnic and other Groups in the United States* (pp. 9-50). Washington, DC: U.S.G.P.O.

77. Plotkin, H. (2003). We-intentionality: an essential element in understanding human culture? *Perspectives in Biology and Medicine,* **46**(2), 283-296.

78. Pratt, A. D. (1977). The information of the image: a model of the communications process. *Libri,* **27**(3), 204-220.

79. Raber, D. (2003). *The problem of information: an introduction to information science*. Lanham, MD: Scarecrow Press.

80. *Random House Unabridged Dictionary* (2nd ed.) (1993). New York, NY: Random House.

81. Rayward, W.B. (1992). Restructuring and mobilising information in documents: a historical perspective. In P. Vakkari & B. Cronin (Eds.), *Conceptions of library and information science: historical, empirical and theoretical perspectives: proceedings of the International Conference Held for the Celebration of 20th Anniversary of the Department of Information Studies, University of Tampere, Finland, 26-28 August, 1991*, (pp. 50-68). London: Taylor Graham.

82. Sandstrom, A.R. & Sandstrom, P.E. (1995). The use and misuse of anthropological methods in library and information science research, *Library Quarterly*, **65**(2): 161-199.

83. Sandstrom, P.E. (1994). An optimal foraging approach to information seeking and use. *Library Quarterly*, **64**(4), 414-449.

84. Sandstrom, P. E. (1999). Scholars as subsistence foragers. *Bulletin of the American Society for Information Science*, **25**(3), 17-20.

85. Sandstrom, P.E. (2001). Scholarly communication as a socioecological system. *Scientometrics*, **51**(3), 573-605.

86. Shannon, C.E. & Weaver, W. (1975). *Mathematical theory of communication.* Urbana, IL: University of Illinois Press.

87. Siegert, R.J. & Ward, T. (2002). Evolutionary psychology: origins and criticisms. *Australian Psychologist*, **37**(1), 20-29.

88. Sperber, D. (1996). *Explaining culture: a naturalistic approach.* Oxford: Blackwell.

89. Stanley, S.M. (1996). *Children of the ice age: how a global catastrophe allowed humans to evolve.* New York, NY: Harmony Books.

90. Stonier, T. (1997). *Information and meaning: an evolutionary perspective.* Berlin: Springer.

91. Stuart-Fox, M. (1999). Evolutionary theory of history. *History & Theory,* **38**(4), 33-51.

92. Sugiyama, M. S. (2001). Narrative theory and function: why evolution matters. *Philosophy and Literature,* **25**(2), 233-250.

93. Talja, S., Keso, H. & Pietiläinen, T. (1999). The production of 'context' in information seeking research: a metatheoretical view. *Information Processing and Management,* **35**(6), 751-763.

94. Talja, S., Tuominen, K. & Savolainen, R. (2005). 'Isms' in information science: constructivism, collectivism and constructionism. *Journal of Documentation,* **61**(1), 79-101.

95. Thompson, F.B. (1968). The organization is the information. *American Documentation,* **19**(3), 305-308.

96. Tuominen, K., Talja, S. & Savolainen, R. (2002). Discourse, cognition and reality: toward a social constructionist metatheory for library and information science. In H. Bruce, R. Fidel, P. Ingwersen, & P. Vakkari (Eds.), *Emerging frameworks and methods: proceedings of the Fourth International Conference on Conceptions of Library and Information Science (CoLIS4), Seattle, WA, USA.* (pp. 271-283). Greenwood Village, CO: Libraries Unlimited.

97. Vakkari, P. & Cronin, B. (Eds.). (1992). *Conceptions of library and information science: historical, empirical and theoretical perspectives: proceedings of the International Conference Held for the Celebration of 20th Anniversary of the Department of Information Studies, University of Tampere, Finland, 26-28 August, 1991*London: Taylor Graham.

98. Waal, F.B.M. de (2002). Evolutionary psychology: The wheat and the chaff. Current directions in *Psychological Science,* **11**(6), 187-191.

99. Wagner, D.G. & Berger, J. (1985). Do sociological theories grow? *American Journal of Sociology,* **90**(4), 697-728.

100. Wellisch, H. (1972). From information science to informatics: a terminological investigation. *Journal of Librarianship*, **4**(3), 157-187.

101. Wersig, G. & Neveling, U. (1975). The phenomena of interest to information science. *The Information Scientist*, **9**(4), 127-140.

102. Wiegand, W. A. (1999). Tunnel vision and blind spots: what the past tells us about the present; reflections on the twentieth century history of American librarianship. *Library Quarterly*, **69**(1), 1-32.

103. Wiener, N. (1961). *Cybernetics: or control and communication in the animal and the machine* (2nd ed.). Cambridge, MA: The MIT Press.

104. Wills, P.R. (1994). Does information acquire meaning naturally? *Berichte der Bunsen-Gesellschaft für Physikalische Chemie*, **98**(9), 1129-1134.

105. Winterhalder, B. & Smith, E.A. (2000). Analyzing adaptive strategies: human behavioral ecology at twenty-five. *Evolutionary Anthropology*, **9**(2), 51-72.

106. Wright, R. (1994). *The moral animal: evolutionary psychology and everyday life.* New York, NY: Vintage Books.

11

METAPHORS OF TIME AND INSTALLED KNOWLEDGE ORGANIZATION SYSTEMS: OUROBOROS, ARCHITECTONICS, OR LACHESIS?

Joseph T. Tennis

University of Washington, Information School, Box 352840 - Mary Gates Hall, Ste. 370, Seattle, WA 98195-2840, USA

ABSTRACT

Introduction

This paper presents three metaphors of time present in knowledge organization systems.

Analysis

These three metaphors the architectonic, ouroboric, and lachesic, can be used as lenses to analyse extant or newly designed knowledge organization systems.

Conclusion

A foundational view of evaluating and theorizing about knowledge organization systems must account for change and time in order for us to take a long view of improving knowledge organization and our understanding of knowledge organization systems, and these metaphors might be helpful. ß

1. INTRODUCTION AND BACKGROUND

The evaluation of knowledge organization systems is the raison d'metre for research library and information science and knowledge organization. Designing and building the ideal knowledge organization system is the motivating factor for critique, fieldwork, and analysis this research and design space. What is considered as the ideal such system is the subject of debates the field and results a variety of different approaches to research (Furner, 2010: 21). Some have argued that if we can represent reality faithfully our knowledge organization systems we will create good systems - or more accurately correct systems. Some examples of this are Bliss and his educational consensus (Bliss, 1933) Hjørland's materialist citation-as-evidence (Hjørland, 1992), and the domain analytic perspective (Mai, 2005; Hjørland and Albrechtsen, 1995).

This trend continues the research literature, with various camps arguing for the correct way to build a correct knowledge organization system (cf. Fox, 2012; Szostak, 2013). Yet, one thing that remains to be answered is does correct hold over a long period of time? And if it does not, how does that effect how we evaluate knowledge organization system? For example, if a domain changes, what happens to the older classification, say 19th century biology? Or theories of race from 1900-Present, (cf. Furner, 2007).

This paper describes current work that examines the effect of time on the semantics of knowledge organization system and then returns to this question of evaluating such systems based on whether or not they are correct. order to do this I will examine a foundational issue: the metaphor used to describe time knowledge organization systems.

It is important to study the effects of time on semantics because we can observe dramatic change large and installed knowledge organization systems, like library classification schemes - changes that effect browsing, retrieval, and sense-making, as well as identify formation and the right to name. These changes are brought about for numerous reasons. There are some social justice reasons, for example, naming a group of people what they want to be called is a reason to change racist language. Other reasons, like changes literary warrant, are also cited. For example, eugenics is a term that was once written about biological sciences, and it is not commonly done so now. So the term eugenics is no longer used the biological sciences of the Dewey Decimal Classification. As a result, long-lived collections, books about the biological aspects of eugenics occupy the same class as books on berries and nuts and other reproductive parts of plants (Tennis, 2012).

With these two examples we can see that the evaluation of (1) a particular knowledge organization system or (2) design requirements for a not-yet-built system has to account for the eventual change semantics over the course of time. If we were correct at one point time, but we are not correct now, what do we do about correcting our mistake? This question requires us to address two things. The first is to ask what is the extent and variety of change extant knowledge organization system? The second is what time metaphor for designing for change might be adopted as an amelioration to the problems presented by the effect of time on knowledge organization systems? We will describe what we currently understand to be the variety and extent of change knowledge organization system first. We will then address the second issue of metaphors.

Time and change, knowledge, literature, or semantics, is a constant concern for the discipline of knowledge organization. E. C. Richardson, his Classification: Theoretical and Practical (1901) describes his laws for classification, specifically the Historical Law and the Law of Evolution. The historical law claims that things, the subject of classification, proceed through space and time from more simple to complex. The corollary to this is the Law of Evolution which states that only those things the grow more complex through time persist, while those that do not grow more complex dissolve and disappear. The recognition of this leads subsequent theorists, like S. R. Ranganathan, to treat the constant change subjects the universe of knowledge as a design constraint. Ranganathan's concern was to build a dynamic theory of classification (Ranganathan revised and edited by Gopinath, 1989 p. 5). This would allow for documents with new subjects to be placed what he called a filiatory sequence (Ranganathan, 1967).

Further, discussions of time and change schemes are a current concern for researchers. Buckland for example has written about subject obsolescence catalogues (Buckland, 2012). this piece he talks about the Janus faced nature of subject description. He says,

The librarian's formal act of naming, of recording the topical description of a document or of specifying a relationship between named topics, is necessarily performed at some point time and inscribed into the apparatus of bibliographies, catalogs, and indexes. As time passes that act recedes from the present into the past. During the same flow of time the prior discourse, upon which the choice of name was derived, has continued, evolved, and changed, and naming practices can be expected to have evolved with those changes. Also, as the future becomes the present, new futures continue to be foreseen, and the forward-looking perspective of the subject cataloger increasingly comes to be related to new and different future discourses. However, an assigned name, once inscribed, is fixed. So, with the passing of time, its relationship with both the then-past discourses and also the then-future expected discourse needs both drift away from the perceptions of an advancing present. Assigned names are, therefore, inherently obsolescent with respect to both the past and the future. Discourses and the librarians flow forward with time, but the assigned names have been inscribed for, and fixed in, a receding past. (Buckland, 2012 p.

Here we see one metaphor for time information systems. This metaphor puts the act of description on a single point the line of time. It also describes the semantics of that description as not being fixed or even relevant at a future or past time. It is one view of change.

Change this case is semantic and is a continuous process that cannot be avoided. It is also a metaphor for time that operates a single dimension - a single line. Others have seen multiple strata operating time and information systems. Boydens and van Hooland analyse databases using a three-strata metaphor for change (2011). The three strata are long-term, medium-term, and short-term changes. This analysis was drawn from Braudel (1949/1996) and used to examine databases. One case they use is social security databases, and that context long-term are rules that change quarterly or yearly, medium-term is the management of the data the databases, and the short-term is the continuously evolving personal interactions citizens have with the system (cf. Tennis, 2010). Given these time scales we see the challenges change can bring. Boydens and van Hooland describe it thus,

Companies regularly merge, split or disappear altogether, while new professions or categories of activity not covered by the official nomenclatures constantly come into existence, as with the diversification of IT jobs. From a dynamic point of view, an ideal database should therefore match the rhythm of its updates to the (unforeseeable) division into 'layered timescales' of the changes reality that it seeks to grasp. We must add to this the necessity, always revealed a posteriori, to integrate unforeseen observations, prohibited a priori by the closed world assumption, which states that, for evident operational reasons, every fact that violates the database formal integrity constraints is considered to be false. (Boydens and van Hooland, 2011: 284).

Boydens and van Hooland posit how we might handle change an information system - that it be constantly updated even a complex multi-strata conception of time scales. This ideal solution solves the problem outlined Buckland (2012) above. If we are constantly updating the database then there is no past or future meaning. However, as they observe, this requires a particular perspective on and analysis of information systems. Likewise the design requirements they offer for information systems to accommodate the need for this constant update of the semantics are specifically creating partial or preliminary and less formal categories. This may only work for some kinds of information systems, and might not work for very formal information retrieval systems.

The problem and the discussion around an amelioration discussed by Boydens and van Hooland seems related to the question of characteristics of infrastructure, outlined by Star (1999) and Star and Ruhleder (1996). Some of those characteristics, like: embeddedness, links with conventions of practice, built on an installed base, becomes visible on breakdown, and is fixed modular increments, not all at once or globally seem related to the Boydens and van Hooland analysis. The difference may be a matter of scale. Perhaps Boydens and van Hooland are talking about a smaller information system infrastructure compared to those discussed by Star and colleagues. Another difference is the metaphor of time used by Star and her collaborators. If a change cannot be fixed globally and all at once, we have a different conception of semantics, what these semantics are linked to, and their relationship to what has come before. This is because time is linked to change knowledge organization systems.

And these relationships between past and present concepts and relationships between concepts are of primary concern for those doing ontogenic analysis of knowledge organization systems. Ontogenic analysis is the description of how a subject changes or a set of classes change a knowledge organization system over time. Major systems like the DDC, UDC, and Wikipedia are the objects of ontogenic analysis (e.g., Salah et al., 2012; Suchecki, et al., and Tennis, 2012). A related area of investigation looks at the versioning of schemes order to propose design recommendations handling change (Hillmann et al., 2010; Tennis, 2007; Tuominen et al., 2011). Much of this research assumes an accretion of meaning and structure over time. this case time is neither a single line, nor is it and constant froth of meaning. It is more geologic or archeological. For example, we can see how different editions of the UDC grow over time. See Figure 1 drawn from Salah et al., (2012).

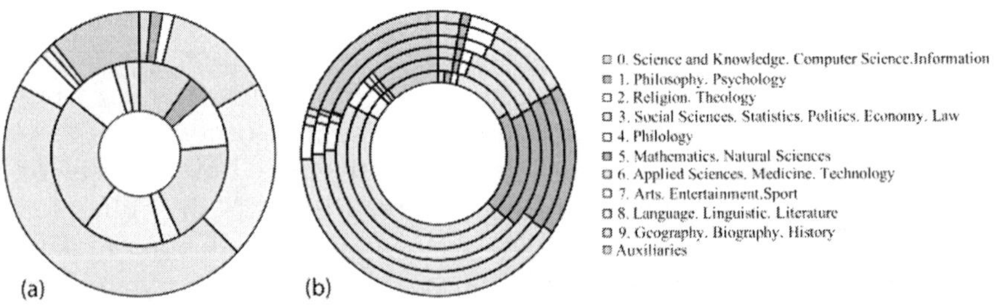

Figure 1. UDC over time. (a) is the distribution of UDC main classes. Inner ring is 1905 and the outer ring is 1994. (b) is the distribution of UDC main classes. 1994 (inner most ring), 1997, 1998, 2005, 2008, and 2009 (outer most ring).

We can also see how a single subject changes position a knowledge organization system, like eugenics the Dewey Decimal Classification. See Figure 2 from Tennis (2012).

Given these two visualizations of ontogenic analysis it seems clear that researchers this field assume a kind of geologic time scale their analyses. This is different form Buckland's conception of time erasing meaning from the past and the prospective future, but perhaps similar, though not identical, to Boydens and van Hooland's strata of time scales and change.

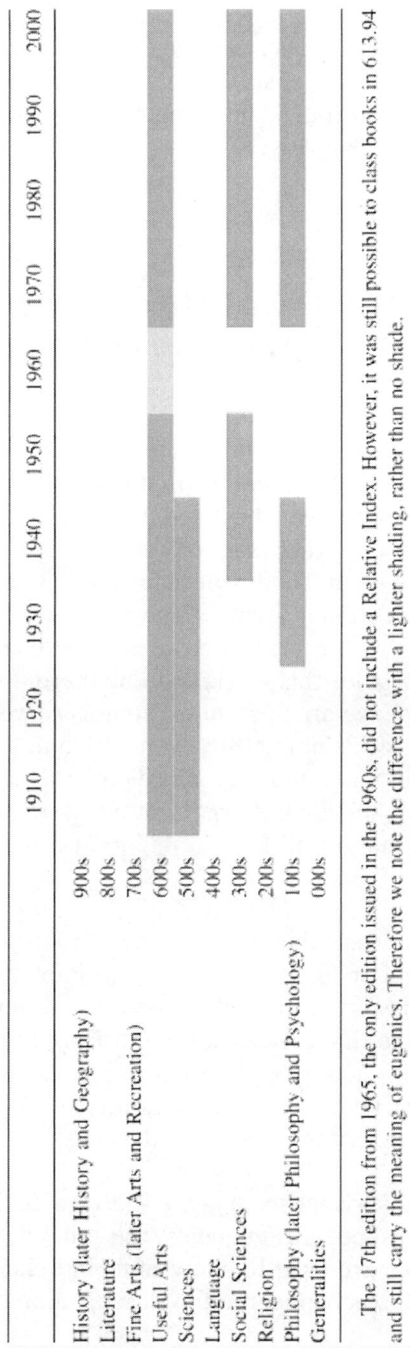

Figure 2. 'Eugenics' the Dewey Decimal Classification organized by decade and displayed at the centuries level of the classification scheme.

Above we have reviewed how time and change are currently conceptualized the literature. We can see that there are perhaps a range of metaphors used, and because of that we have still to address the question of evaluation through these metaphors. Now we can address those metaphors more explicitly.

2. METAPHORS OF TIME AND CHANGE KNOWLEDGE ORGANIZATION SYSTEMS

Above we described some of the current ways which researchers knowledge organization have studied how things change knowledge organization systems. We have also started to point out where metaphors of time and change have been invoked by researchers, either explicitly, or implicitly. This section we address some of those metaphors and name them. The purpose for doing this is so that we can identify the lens through which we can evaluate knowledge organization systems over time. It is clear that there will be debate over which metaphor is best for this work, just like there is debate on how to define when a knowledge organization system is correct, and when it is not (see above). The value doing this, even with a debate anticipated, is that we gain deeper understanding, a more fundamental understanding, of our evaluation of knowledge organization systems by being more explicit about the kinds of analysis possible and desirable to do this work.

The ontogenic analysis above, we saw a kind of geologic or archaeological metaphor for time and change knowledge organization system. Each edition was considered a different stratum, independent of its publication date. So that the case of the UDC we see that 1994, 1997, 1998, 2005, 2008, and 2009 are all visualized as the same intervals and volumes, the only variation being the quantity of main classes., which changes the distribution of the colours of the rings.

Likewise, the analysis visualized Figure 2 above considers either the presence or the absence of eugenics the knowledge organization system. Here decades are used to show time, but change is marked by whether or not the subject appeared the classification schedules.

Both of these examples of ontogenic analysis operate on an architectonic metaphor for time and change. This sense architectonic means systematic, designed, and formalized meaning progressing through time. This metaphor assumes that meaning is discrete from edition to edition, and that knowledge organization system are potentially ruptured through the change that results a new edition. We can contrast this with more incremental architectonic time where we do not issue editions, but rather revise piecemeal one concept at a time, without the declaration of a new edition. This has been observed the theoretical literature (Tennis and Sutton, 2008) as well as practice the form of the Library of Congress weekly updates (Library of Congress, 2013). Both of these kinds of architectonic metaphors are different from what Buckland and Boydens and van Hooland use.

Both papers address above, the Buckland and the Boydens and van Hooland, consider change meaning to be immediate, even though it is tied to the past and directs meaning for future use knowledge organization systems For Buckland the subject description assigned by librarians is constantly moving toward obsolescence. As he says, "an assigned name, once inscribed, is fixed. So, with the passing of time, its relationship with both the then-past discourses and also the then-future expected discourse needs both drift away from the perceptions of an advancing present," (Buckland, 2012 p. 284). For Boydens and van Hooland the interaction of users with databases should signal an immediate change meaning even negotiation with a longer time of juridical change (long-term), database management time (medium-term). both of these cases we can see that change is a new meaning, springing forth from an old meaning, but with no guarantee for fixity the future. So meaning is recreated with a tie to the past. Metaphorically this is like the ouroboros, the serpent eating its own tail. See Figure 3. This metaphor implies that not only is something new constantly emerging from old, and that it too will end. It also signifies that this process of almost immediate semantic shift will never end. This is not unlike Mai's description of unlimited semiosis (Mai, 2001). The question that does surface when is it practicable to consider time this way knowledge organization systems? When is it ever possible to be correct if we never have a fixed meaning?

This metaphor of constant renewal has a corollary the practice of knowledge organization. The editors of the Dewey Decimal Classification system have, on occasion, phoenixed particular sections of the schedules. That is, they have erased one set of classes and their hierarchy, and replaced it with a new set of classes. This is a structural-semantic change that happens on a longer time scale than invoked by either Buckland or Boydens and van Hooland. The act of phoenixing is part why the architectonic metaphor is used when considering editions of knowledge organization systems.

Figure 3. Drawing by Theodoros Pelecanos, the alchemical tract Synosius (1478).

There is one final metaphor that we want to consider here. That is of time knowledge organization systems as an allotted amount, from beginning to end. this metaphor change has a singular relationship with what has come before, it is additive. this metaphor, invoked by both Richardson and Ranganathan, knowledge at any one point time can be represented, but if we need to add we can, a single direction from past to present, by adding to the system. This latter was a primary concern for Ranganathan, whose dynamic theory of classification sought to establish how new facets and isolates could be added to schedules for classification without disrupting the order established by earlier editions of the scheme.

This metaphor we call lachesic after the Greek and Roman goddess Lachesis, the Fate that decides the lot of a human's life, from birth to death. She measures out the entirety of the experience. Lachesis is her Greek name.

Ranganathan's desire was to create a single line of knowledge organization, by allowing new subjects to emerge and be accepted hospitably into the extant scheme. There are potential problems with this, because it assumes that the meaning of a single subject stays the same even with the addition of new subjects. For example, engineering when it first appears the Colon Classification is the engineering of the early twentieth century. Is it the same as the engineering of today? Does the extension of engineering change with the addition of nano-engineering as new subject? We are starting to study this particular problem, but there are many aspects to it that should be considered (Tennis, 2012).

Each of these metaphors, the architectonic, ouroboric, and lachesic, govern our views on time and change knowledge organization system. the case of the ouroboric model of time and change knowledge organization system, we have to decide when and how we understand a particular knowledge organization system to be correct. If we follow domain analysis, which situates validation of correct knowledge organization system the corroboration between domain and the particular knowledge organization system, we should be able to ascertain some fixity of meaning. Otherwise we would not be able to build a knowledge organization system from the analysis of a domain if the semantics were constantly shifting. This means that domain analysis cannot align domain analysis with this metaphor of time and change knowledge organization system.

However, we might be able to read Buckland as concluding that the ouroboric metaphor is not immediate, that is progresses slowly. this case, we are a constant state of validation between domain and knowledge organization system. This is the manifestation of the snake eating its tail. The knowledge organization system is built according to the domain analysis, and then the subject description begins to fade into obsolescence, so the need to assess the domain resurfaces, and the cycle begins again.

The lachesic metaphor of time we assume that once knowledge organization system is correct its only change is to add concepts and relationships that did not exist before. There is no commitment evidenced the literature that practitioners must use older concepts.

Having said that, older concepts are not revised the lachesic models of change knowledge organization system. Domain analysis a lachesic model would only introduce new concepts and relationships. It would not revise old concepts. Again, we have to ask the question as to how often, one would reconsider the domain a lachesic model.

Architectonic models of change and time knowledge organization system are, a way, the most measured of the time metaphors we have so far discussed. The only time we consider when we should consult the domain, if domain analysis is our modus operandi, is when we consider issuing a new edition. Likewise, we only build relationships between what has been revised and what has not when we issue a new edition the architectonic metaphor. The incremental architectonic metaphor we might see the reconsideration of the domain move more quickly, but this is, as far as I know, currently an unknown.

Each of these three metaphors can be loosely aligned with perspectives on indexing, which dominate the literature. the case of architectonic metaphor, the primacy seems to be placed on the system and its changes. With the ouroboros metaphor see the user's meaning surface relation to change and time. With the lachesic metaphor we perhaps see documents and their semantics given primacy to change the scheme by adding more over time. These are rough approximations, but they could be useful comparisons to make.

This section we have described three metaphors for conceptualizing time and change the context of knowledge organization system. These three metaphors were our interpretation of time the context of a subset of extant studies. If we see these metaphors at play, then what effect does time have and our appending time through a particular metaphor on knowledge organization systems?

3. EFFECT OF TIME ON KNOWLEDGE ORGANIZATION SYSTEMS

The effect of time on knowledge organization systems (knowledge organization system) is measured by the metaphor used to describe change. If an edition is published 1876, but no class or words associated with a class, nor any relationship is changed, then there is no time considered. The only caveat here would be the use of the class for a particular collection. For example, if a single class contained books on a topic, but the interpretation of that word and class (and its relationships) was reinterpreted by classing books on other topics the same class over time, then we would have some change - change extension of books for the class over time. The lachesic metaphor of time might lead us to something like this. Leaving this very interesting hypothetical (but possibly observable) caveat aside we can operationalize our definition of time as change knowledge organization system.

This is an important operationalization evaluating knowledge organization system, because changing a knowledge organization system might mean it is no longer correct. Or the knowledge organization system is changed because it is no longer correct. This sense of motion, moving from incorrect to correct through change the knowledge organization system influences our discussions of evaluation of knowledge organization system, for none of them seem impervious to change. Yet we must understand what change does to knowledge organization system. That is, is there any sacrifice for making a knowledge organization system correct through change?

4. EVALUATING KNOWLEDGE ORGANIZATION SYSTEMS OVER TIME

The degree to which we currently evaluate information systems is by their degree of correctness. While this is not the only way to evaluate knowledge organization systems, it is the dominant one, and shapes much of the discourse and research knowledge organization. Given this practice, and given that we might employ different metaphors for time and change these systems, we can beg to question the atemporality of current discussions of knowledge organization systems. If a system, at this point time, comports with the current educational consensus (following Bliss's conception), does that mean it always has and always will? When will it comport if things change? When will it not comport? Further, what is the relationship between retrieval, browsing, and sense-making large installed system when they do undergo change? Or how do we help users whose semantics may change from day-to-day as presaged by Buckland and Boydens and van Hooland? The past is prologue for the evaluation of knowledge organization systems.

5. CAVEAT AND CONCLUSION

It is possible to change our assumptions about how we evaluate knowledge organization systems. We may not follow the dominant (yet diverse) paradigm of evaluating such systems based on a level or degree of correctness. Following Feinberg, it may be possible to consider more rhetorical or artistic issues (Feinberg, 2010a and2010b). We might also consider aesthetic issues (Ojennus and Tennis 2013a and 2013b). these cases time would perhaps not be operationalized as change, though it could be. And the evaluation of knowledge organization systems built on rhetorical, artistic, or aesthetic principles rather than principles that are meant to lead to a correct representation, would likely use different criteria, for example, how does the knowledge organization system comport with some aesthetic ideal? This is not a concern for those who want a system to corroborate the correct way of representing knowledge.

So with this consideration, we might conclude that a significant, if not foundational issue for the evaluation of knowledge organization systems is to account for change over time, and specifically, the methodological and philosophical concerns that grow out of particular metaphors for change such systems. It perhaps is time for us to take the long view on how we evaluate knowledge organization systems, because change will come.

REFERENCES

1. Aalberg, T. (2002). Navigating bibliographic catalogues. *Proceedings of the 6th European Conference on Research and Advanced Technology for Digital Libraries ECDL 2002*, **LNCS 2458** (pp. 238-250). Berlin: Springer.

2. American Association of Law Libraries. (2010). Competencies of Law Librarianship. *Retrieved 28 February, 2013 from http://* www.aallnet.org/main-menu/Leadership-Governance/policies/ *PublicPolicies/competencies.html*

3. American Association of School Librarians. (2010). Standards for Initial Preparation of School Librarians. *American Library*

Association. Retrieved 28 February, 2013 from https://lita.org/ala/mgrps/divs/aasl/aasleducation/schoollibrary/2010_standards_and_items_with_statements_of_scope.pdf

4. American Library Association. (2009). Core Competences of Librarianship. *Retrieved 28 February, 2013 from* http://www.*ala.org/educationcareers/careers/corecomp/corecompetences*

5. Bliss, H. E. (1933). The organization of knowledge libraries and the subject-approach to books. *Wilson: New York.*

6. Boydens, I. & van Hooland, S. (2011). "Hermeneutics Applied to the Quality of Empirical Databases." *Journal of Documentation* **67**(2): 279-289.

7. Braudel, F. (1949/1996). La Méditerranée et le Monde Méditerranéen à l'Epoque de Philippe II. *Three volumes. (University of California Press).*

8. Buckland, M. (2012). "Obsolescence Subject Description" *Journal of Documentation* **68**(2): 154-161.

9. Feinberg, M. (2010a). "Two Kinds of Evidence: How Information Systems Form Rhetorical Arguments." *Journal of Documentation* **66**(4): 491-512.

10. Feinberg, M. (2010b). "Designing Collecitons for Sotryteling: Purpose, Pathos, and Poetry." *Information Research*, **15**(3). Paper CoLIS 701. Available: Available at http://InformationR.net/ir/15-1/colis7/colis701.html

11. Fox, M. (2012). Review of Szostak, Rick. 2003. A Schema for Unifying Human Science: Interdisciplinary Perspectives on Culture. Selinsgrove, PA: Susquehanna UP And Szostak, Rick. 2004. Classifying science: Phenomena, Data, Theory, Method, Practice. Norwell, MA: Springer. *Knowledge Organization* **39**(4): 300-303.

12. Furner, J. (2007) "Dewey Deracialized: A Critical Race-Theoretic Perspective." *Knowledge Organization* **34**(3): 144-168.

13. Furner, J. (2010). "Philosophy and Information Studies." Blaise Croned. *Annual Review of Information Science and Technology*, vol. 44. (Information Today).

14. Hillmann, D., Coyle, K., Phipps, J. & Dunsire, G. (2010). RDA Vocabularies: Process, Outcome, and Use. *D-Lib Magazine* **16**(1/2). Available: http://dlib.org/dlib/january10/hillmann/01hillmann.html

15. Hjørland, B. & Albrechtsen, H. (1995). "Toward a New Horizon Information Science: Domain-analysis." *Journal of the American Society for Information Science* **46**(6): 400-425.

16. Hjørland, B. (1992). "The Concept of Subject Information Science." *Journal of Documentation* **48**(2): 172-200.

17. Library of Congress. (2013). Library of Congress Subject Headings (LCSH) Approved Lists. *Available:* http://www.loc.gov/aba/*cataloging/subject/weeklylists/*

18. Mai, J-E., (2005). "Analysis Indexing: Document and DomaCentered Approaches." *Information Processing and Management* **41**(3): 599-611.

19. Mai, J-E., (2001). "Semiotics and Indexing: An Analysis of the Subject Indexing Process." *Journal of Documentation* **57**(5): 591-622.

20. Ojennus, P. & Tennis, J. T. (2013). "Modeling the aesthetic axis of information organization frameworks, part 1: Theoretical basis."*Journal of Documentation*. **69**(6): TBA.

21. Ojennus, P. & Tennis, J. T. (2013). "Modeling the aesthetic axis of information organization frameworks, part 2: Case studies." *Journal of Documentation*. **69**(6): TBA.

22. Pelecanos, T. (1478). Fol. 196 of Codex Parisinus graecus 2327, a copy (made by Theodoros Pelecanos (Pelekanos) of Corfu Khandak, Iraklio, Crete 1478) of a lost manuscript of an early medieval tract which was attributed to Synosius (Synesius) of Cyrene (d. 412).

23. Ranganathan, S. R. revised and edited by Gopinath, M. A. (1989). Colon Classification edition 7. (Sarada Ranganathan Endowment for Library Science: Bangalore).

24. Ranganathan, S. R. (1967). *Prolegomena to library classification.* Bombay: Asia Publishing House

25. Richardson, E. C. (1901). *Classification theoretical and practical.* New York, NY: Charles Scribner's Sons.

26. Salah, A.A.A., Gao, C., Suchecki, K., Scharnhorst, A. & Smiraglia, R. P. (2012). "The evolution of classification systems: Ontogeny of the UDC." Preprint on Arxiv http://arxiv.org/abs/1204.3769 Forthcoming P*roceedings of ISKO* 2012.

27. Star, S.L. (1999). "The Ethnography of Infrastructure." *American Behavioral Scientist* **43**(3): 377-391.

28. Star, S.L. & Ruhleder, K. (1996). "Steps toward an ecology of infrastructure: Design and access for large information spaces."*Information Systems Research* **7**(1): 111-134.

29. Szostak, R. (2013). "Speaking Truth to Power Classification: Response to Fox's Review of My Work, KO 39:4, 300." *Knowledge Organization* **40**(1): 76.

30. Tennis, J.T. (2012). "The Strange Case of Eugenics: A Subject's Ontogeny a Long-Lived Classification Scheme and the Question of Collocative Integrity." *Journal of the American Society for Information Science and Technology* **63**(7): 1350-1359.

31. Tennis, J.T. (2012). "Facets and Fugit Tempus: Considering Time Faceted Classification Schemes." (2012). *Proceedings of the 12th International Conference for Knowledge Organization*. (Mysore, India). Advances Knowledge Organization vol. **13**. Ergon: Würzburg: 58-62.

32. Tennis, J.T. (2010). "Measured Time: Imposing a Temporal Metric to Classificatory Structures." *Proceedings of the 11th International*

Conference for Knowledge Organization. (Rome, Italy). Advances Knowledge Organization vol. **12**. Ergon: Würzburg: 223-228.

33. Tennis, J.T. & Sutton, S. A. (2008). “Extending the Simple Knowledge Organization System (Sknowledge organization system) for Concept Management Vocabulary Development Applications.” *Journal of the American Society for Information Science and Technology*. **59**(1): 25-37.

34. Tennis, J.T. (2007). "Scheme Versioning the Semantic Web." *Cataloging and Classification Quarterly* **43**(4/3): 85-104.

35. Tuominem, J., Laurenne, N. & Hyvönen, E. (2011). Biological Names and Taxonomies on the Semantic Web - Managing the Change Scientific Conception. *Proceedings of Extended Semantic Web Conference 2011*. Available: www.seco.tkk.fi/publications/2011/tuominen-et-al-taxmeon-2011.pdf

INDEX

R

S

T